ABSTRACT ALGEBRA

A GENTLE INTRODUCTION

TEXTBOOKS in MATHEMATICS

Series Editors: Al Boggess and Ken Rosen

TEXTBOOKS in MATHEMATICS

ABSTRACT ALGEBRA

A GENTLE INTRODUCTION

Gary L. Mullen

The Pennsylvania State University
University Park, USA

James A. Sellers

The Pennsylvania State University
University Park, USA

CRC Press
Taylor & Francis Group
Boca Raton London New York

CRC Press is an imprint of the
Taylor & Francis Group an **informa** business

A CHAPMAN & HALL BOOK

CRC Press
Taylor & Francis Group
6000 Broken Sound Parkway NW, Suite 300
Boca Raton, FL 33487-2742

First issued in paperback 2022

Version Date: 20161121

ISBN 13: 978-1-03-247697-1 (pbk)
ISBN 13: 978-1-4822-5006-0 (hbk)
ISBN 13: 978-1-315-37260-0 (ebk)

DOI: 10.1201/9781315372600

Visit the Taylor & Francis Web site at
http://www.taylorandfrancis.com

and the CRC Press Web site at
http://www.crcpress.com

Contents

Preface

The aim of this textbook is to provide a brief introduction to abstract algebra. We use the term "A Gentle Introduction" because we do not go into great depth while covering various topics in abstract algebra. Instead, we provide the reader with coverage of numerous algebraic topics to give the reader a flavor of what we believe to be the most important areas of abstract algebra.

For example, we do not cover topics such as the Sylow Theorems in group theory. Instead, we lead the reader through a proof of Lagrange's Theorem, and then in Chapter 8, we provide a beautiful application of this famous group theory result to the study of error-correcting codes used so frequently in today's modern communications.

We also discuss the famous RSA cryptographic system and the Diffie/Hellman key exchange for the secure transmission of information. In addition, we discuss some fascinating ideas in the study of sets of mutually orthogonal latin squares, an area of study that dates back to Leonard Euler in 1782.

The text is intended to be used for a one-semester course in abstract algebra. It is suitable for an introductory course in abstract algebra for mathematics majors. The text is also very suitable for education majors who need to have an introduction to various topics in abstract algebra.

Theorems, definitions, examples, corollaries, etc., are all numbered consecutively within each chapter. For example, in the first chapter, item 1.3 is Theorem 1.3, item 1.5 is Example 1.5, and item 1.8 is Corollary 1.8.

In Chapter 9, entitled "Appendix," we provided a brief review of numerous topics with which the student may not be familiar. This chapter is also meant for the students who need to refresh their memory on a few topics. The topics covered in the Appendix include mathematical induction, the well-ordering principle, sets, functions, permutations, matrices, and complex numbers.

Numerous exercises are provided at the end of almost every section. In addition, in Chapter 10, entitled "Hints and Partial Solutions to Selected Exercises," we provide solutions to the odd-numbered problems, leaving the even-numbered problems for homework use by the instructor. Exercises are numbered using the system Exercise i.j.k. This refers to the k-th exercise in Chapter i, Section j.

We would like to thank Serge Ballif for his help in setting up the LaTeX files for our manuscript. Gary Mullen would also like to thank his wife (Bev Mullen) for her patience and understanding during the writing of this text. Thanks are also due the staff at CRC Press. Finally Gary Mullen would like to thank his Math 470 (Algebra for Teachers) class held during the Spring Semester 2016 at the Pennsylvania State University for their help in locating errors, typos, and in general suggesting a number of clarifications and improvements in the text.

    G.L. Mullen

    J.A. Sellers

Chapter 1

Elementary Number Theory

This chapter deals with several fundamental topics from elementary number theory. The reader will find that many of these topics will later be generalized and discussed in other chapters such as the group, ring, and field chapters.

1.1 Divisibility

We begin by discussing the notion of divisibility of integers. Indeed, this is one of the oldest concepts in number theory. The topic has been studied in depth for at least 2000 years. In fact, Euclid ($\sim$325BC–265BC) dealt with the topic of divisibility in Book VII of his *Elements*.

Let a be an integer and let d be an integer. We say that d **divides** a if there is an integer k so that $a = dk$. The important point here is that k must be an integer. If there are such integers, we denote the fact that d divides a by using the notation $d|a$.

For example, $2|8$ since $8 = 2(4)$; $36|108$ since $108 = 36(3)$; $3|(-36)$ since $-36 = 3(-12)$; and for any integer m, $3|(15m + 3)$ since $15m + 3 = 3(5m + 1)$. On the other hand, 3 does not divide 13 as there is no integer k with $13 = 3k$.

The reader should be aware that $d|a$ is just a notation for the fact that the integer d divides a. Be careful, do not confuse this with the notation d/a which, as usual, represents the value d divided by a.

In the following lemma, we provide a few basic properties involving the divisibility of integers.

Lemma 1.1 *Let a, b, d be integers with $d > 0$.*

- *(i) If $d|a$ and $d|b$, then $d|(a + b)$;*

- *(ii) If $d|a$ and $d|b$, then $d|(a - b)$;*

- *(iii) If $d|a$ and $d|b$, then $d|(ma + nb)$ for any integers m and n;*

- *(iv) If $m|d$ and $d|a$, then $m|a$.*

Proof: To prove part (i), we may assume that $a = dk$ and $b = dl$ where k and l are integers. Hence $a + b = dk + dl = d(k + l)$ so that d divides $a + b$ (since $k + l$ is an integer).

The proof of part (ii) is similar and hence omitted. Proofs of the remaining parts are left to the exercises. ∎

We now state two extremely important results about positive integers. These results are known as the Division and Euclidean Algorithms, respectively. The reader should not be confused or worried about the term "algorithm;" it is simply a term that is used to indicate a method to compute something. While mathematicians often use this terminology, it is also very heavily used by people working in the field of computer science.

Theorem 1.2 *(Division Algorithm) Let a and b be two positive integers. Then there are integers q and r so that $b = aq + r$ where $0 \leq r < a$.*

A proof of this result is given in the Appendix, Section 2 where we discuss the Well-Ordering Principle. For the moment, let's just use this result.

The integers q and r are often called the **quotient** and the **remainder**, respectively.

This terminology comes from long division. For example, consider the integers $a = 4$ and $b = 23$. Dividing 23 by 4, we see that $23 = 5(4) + 3$. Hence from the Division Algorithm the quotient q is 5 and the remainder r is 3.

Note in the Division Algorithm that $0 \leq r < a$. As in the process of long division, this simply says that the remainder is always non-negative and less than the divisor.

Given two positive integers a and b, we define the **greatest common divisor** of a and b to be the largest positive integer that divides both a and b. This is often denoted by writing $\gcd(a, b)$ or sometimes by simply writing (a, b). With this latter notation one must be careful not to get this confused with the open interval (a, b) or the point (a, b) in the usual xy-plane!

We say that two positive integers are **relatively prime** if their greatest common divisor is 1; i.e., if for integers a and b, $\gcd(a, b) = 1$. Thus two integers a and b are relatively prime if the only factor they have in common is 1. For example, 5 and 16 are relatively prime as are 24 and 37. However, 6 and 28 are not relatively prime since they have a common factor of 2. You should convince yourself that a prime p is relatively prime to any integer n as long as p does not divide the integer n.

How do we find or compute the greatest common divisor of two positive integers a and b, especially if the integers a and b are large? The Euclidean Algorithm does the trick for us by providing a very systematic and efficient way to calculate our greatest common divisor. Euclid demonstrated amazing insight when he devised this algorithm and noted it in Book VII of his *Elements*.

We now state the Euclidean Algorithm for finding the greatest common divisor $\gcd(a, b)$ of two positive integers a and b.

Theorem 1.3 *(Euclidean Algorithm) Let a and b be positive integers. If a divides b then the $\gcd(a, b) = a$. Otherwise, there exists a strictly decreasing sequence of positive integers $r_1, \ldots, r_n$ so that*

$$
\begin{aligned}
b &= aq_1 + r_1 \\
a &= r_1 q_2 + r_2 \\
r_1 &= r_2 q_3 + r_3 \\
&\vdots \\
r_{n-2} &= r_{n-1} q_n + r_n \\
r_{n-1} &= r_n q_{n+1} + 0.
\end{aligned}
$$

Then $\gcd(a, b) = r_n$.

The proof of the Euclidean Algorithm relies heavily on the Division Algorithm (Theorem 1.2). We will omit the proof and instead illustrate the algorithm with several examples. Basically, the algorithm says to repeatedly apply the Division Algorithm over and over until we can't divide any longer. Since the integers a and b are positive, the process must stop because the remainders keep getting smaller each time we apply the Division Algorithm.

Example 1.4 *By way of illustration, we use the Euclidean Algorithm to find the greatest common divisor of $a = 18$ and $b = 58$. By repeated use of the Division Algorithm we have*

$$
\begin{aligned}
58 &= 18(3) + 4 \\
18 &= 4(4) + 2 \\
4 &= 2(2) + 0.
\end{aligned}
$$

Thus, since the last non-zero remainder is 2, $\gcd(18, 58) = 2$.

By using the Euclidean Algorithm in reverse, we can obtain integers x and y so that $\gcd(a, b) = ax + by$. This will prove to be useful in our later work.

As an illustration, consider our previous example. From the next to last line, we note that $\gcd(18, 58) = 2 = 18 - 4(4)$. Then using the line above we have

$$2 = 18 - 4(58 - 3(18)) = 18(13) - 58(4)$$

so that $x = 13$ and $y = -4$.

Example 1.5 *As another illustration of the Euclidean Algorithm, let $a = 68$ and $b = 249$. Then repeatedly using the Division Algorithm we obtain:*

$$
\begin{aligned}
249 &= 68(3) + 45 \\
68 &= 45(1) + 23 \\
45 &= 23(1) + 22 \\
23 &= 22(1) + 1 \\
22 &= 1(22) + 0.
\end{aligned}
$$

Thus, $\gcd(68, 249) = 1$. This means 68 and 249 are relatively prime. We now illustrate one method to find integers x and y so that

$$1 = 68x + 249y.$$

Using the Euclidean Algorithm in reverse, we first note from the next to last line that $1 = 23 - 22$. Then from the line above we have that

$$1 = 23 - 1(45 - 1(23)) = 2(23) - 1(45).$$

Continuing, we obtain

$$1 = 2(68 - 1(45)) - 1(45) = 2(68) - 3(45)$$

and finally using the first line of the above example, we have

$$1 = 2(68) - 3(249 - 3(68)) = 68(11) - 249(3).$$

Hence $x = 11$ and $y = -3$.

We take the liberty of pointing out now that similar results hold for polynomials; we will study these polynomial versions of the Division and Euclidean Algorithms in more detail in Chapter 7.

1.1 Exercises

1. Let m and k be positive integers. Which of the following hold?

   - $5 | 635$

- $-5|635$
- $48|124$
- $341|32871$
- $5|(15m - 10)$
- $m|(-3m)$
- $(k + m)|(7k + 14m)$
- $k|(-6k^2 - k)$

2. Prove part (iii) of Lemma 1.1.

3. Prove part (iv) of Lemma 1.1.

4. Find $\gcd(35, 180)$.

5. Find $\gcd(35, 225)$.

6. Find $\gcd(224, 468)$.

7. Find $\gcd(384, 434)$.

8. If $a = 140$ and $b = 146$, find the $\gcd(a, b)$ and write $\gcd(a, b)$ as $140r + 146s$ for some integers r and s.

9. The sequence of numbers $1, 1, 2, 3, 5, 8, 13, 21, \ldots$ is known as the sequence of **Fibonacci numbers**. After the first two values, a given number is obtained as the sum of the previous two numbers. We denote this sequence by $F_1, F_2, F_3, \ldots$, in honor of Fibonacci who first wrote about these numbers in his book *Liber Abaci*, which was published in 1202.

   Show that, for any $k \geq 1$, $\gcd(F_k, F_{k+1}) = 1$, i.e., show that any two consecutive Fibonacci numbers are relatively prime.

10. If a positive integer d divides the integer a and d also divides the integer b, show that d divides $ar - bs$ for all integers r and s.

11. Show that, for any positive integer n, 3 divides $4^n - 1$.

12. Is $n^3 - n$ divisible by 6 for each positive integer n? If so, show it, and if not, find an example where it fails.

13. For a positive integer a, what are the possibilities for the quantity $\gcd(a + 3, a)$? Find specific examples to demonstrate each possibility.

1.2   Primes and factorization

Prime numbers are the building blocks of the integers. Primes have fascinated mathematicians for thousands of years. We begin by first carefully defining what is meant by a prime number.

A positive integer p is a **prime** if p has exactly two distinct positive divisors, namely 1 and p itself. Thus $2, 3, 5, 7, 11$ are the five smallest primes. Note that positive integers like $4, 9, 16, 18$, and 25 are not primes. Note also that 2 is the only even prime.

The reader should note that with this definition the positive integer 1 is not a prime, since it does not have exactly two distinct divisors. Its only divisor is 1. Early mathematicians may have viewed 1 as a prime, but once we discuss the theory of unique factorization of integers, we will see why it is important not to count 1 as a prime.

How can one test whether a given positive integer n is a prime? This, in general, is not an easy problem, especially when n is large. For example, is 4999 a prime? It turns out that 4999 is indeed a prime, but 4997 is not a prime, though upon quick inspection, it is not so easy to determine these facts.

How does one list all of the primes up to some value n? There is a way to do this, known as the **Sieve of Eratosthenes**, named in honor of Eratosthenes (276BC–194BC), who appears to be the first to make note of this process. The process is quite efficient as long as n isn't "too large."

One begins by listing all of the numbers from 2 to n. Then since 2 is a prime, we leave it in the list and delete all multiples of 2 (except 2 itself) up to n. We thus delete all even values larger than 2. We then leave 3 and delete all larger multiples of 3. The next value not already deleted is 5, so we leave it and delete all multiples of 5. We continue this process with 7, which is yet to be deleted, then 11, etc. The values remaining in the list give all primes up to n.

We now illustrate the Sieve of Eratosthenes by finding all primes up to $n = 50$. We begin by listing all of the positive integers from 2 through 50.

	2	3	4	5	6	7	8	9	10
11	12	13	14	15	16	17	18	19	20
21	22	23	24	25	26	27	28	29	30
31	32	33	34	35	36	37	38	39	40
41	42	43	44	45	46	47	48	49	50

We boldface the number 2 as the first item in this list (since we know it is prime), and then cross out each multiple of 2 that is greater than 2. It is important to note that no actual arithmetic must be done here! We simply start at 2, skip by the amount of 2 (which gets us to the number 4), cross out the 4, then skip by another 2 to get to 6, cross out the 6, and so on. This stage of the process is quite straightforward. This now leaves us with the following table.

2	3	~~4~~	5	~~6~~	7	~~8~~	9	~~10~~
11	~~12~~	13	~~14~~	15	~~16~~	17	~~18~~	19	~~20~~
21	~~22~~	23	~~24~~	25	~~26~~	27	~~28~~	29	~~30~~
31	~~32~~	33	~~34~~	35	~~36~~	37	~~38~~	39	~~40~~
41	~~42~~	43	~~44~~	45	~~46~~	47	~~48~~	49	~~50~~

Once we have traversed the entire list, we then return to the beginning of the list and look for the first number that has not be selected as a prime already and that has not been crossed out. At this stage, that number is 3. We are guaranteed that this is a prime, so we boldface that number and then cross out all multiples of 3 in the list (again by simply skipping by 3 each time and crossing out the corresponding numbers). That leaves us with the following:

2	**3**	~~4~~	5	~~6~~	7	~~8~~	~~9~~	~~10~~
11	~~12~~	13	~~14~~	~~15~~	~~16~~	17	~~18~~	19	~~20~~
~~21~~	~~22~~	23	~~24~~	25	~~26~~	~~27~~	~~28~~	29	~~30~~
31	~~32~~	~~33~~	~~34~~	35	~~36~~	37	~~38~~	~~39~~	~~40~~
41	~~42~~	43	~~44~~	~~45~~	~~46~~	47	~~48~~	49	~~50~~

Returning to the beginning of the list, we see that 5 is the first number which is neither boldfaced nor crossed out. So it must be a new prime. We boldface it and then cross out the multiples of 5 as described in the previous steps.

2	**3**	~~4~~	**5**	~~6~~	7	~~8~~	~~9~~	~~10~~
11	~~12~~	13	~~14~~	~~15~~	~~16~~	17	~~18~~	19	~~20~~
~~21~~	~~22~~	23	~~24~~	~~25~~	~~26~~	~~27~~	~~28~~	29	~~30~~
31	~~32~~	~~33~~	~~34~~	~~35~~	~~36~~	37	~~38~~	~~39~~	~~40~~
41	~~42~~	43	~~44~~	~~45~~	~~46~~	47	~~48~~	49	~~50~~

Once we have fully completed the sieving process, our list will look like this:

2	**3**	~~4~~	**5**	~~6~~	**7**	~~8~~	~~9~~	~~10~~
11	~~12~~	**13**	~~14~~	~~15~~	~~16~~	**17**	~~18~~	**19**	~~20~~
~~21~~	~~22~~	**23**	~~24~~	~~25~~	~~26~~	**27**	~~28~~	**29**	~~30~~
31	~~32~~	~~33~~	~~34~~	~~35~~	~~36~~	**37**	~~38~~	~~39~~	~~40~~
41	~~42~~	**43**	~~44~~	~~45~~	~~46~~	**47**	~~48~~	~~49~~	~~50~~

We find that the list of all primes up to 50 is given by

$$2, 3, 5, 7, 11, 13, 17, 19, 23, 29, 31, 37, 41, 43, 47.$$

In Exercise 1.2.1 below the reader will be asked to list all of the primes up to 100.

We now wish to identify how the primes serve as the "building blocks" of the integers. To do so, we must develop a number of additional results on divisibility. The first is the following theorem:

Theorem 1.6 *Let a, b, c be positive integers with a and b relatively prime.*
   (i) If $a|bc$ then $a|c$;
   (ii) If $a|c$ and $b|c$, then $ab|c$.

Proof: We first prove part (i). Since a and b are relatively prime, we know $\gcd(a, b) = 1$, so there are integers r and s so that $1 = ar + bs$. Hence, $c = car + cbs$. Recall that if an integer divides two integers, it must divide the sum and difference of those two integers. Thus since a divides the product bc and a certainly divides car, it must also divide the right side of the equation $c = car + cbs$. Therefore, a must divide c.

For part (ii), since a divides c, ab divides cbs and since b divides c, ab divides car. Thus ab must divide the sum, which is c. ∎

Our next theorem is often attributed to Euclid (and is referred to by many as "Euclid's Lemma.") It provides the foundation for the truly important role that the primes play in elementary number theory.

Theorem 1.7 *If p is a prime and p divides ab, then p divides a or p divides b.*

Proof: Since p is a prime, we know $\gcd(p, a)$ is either 1 or p. In the latter case p divides a. If $\gcd(p, a) = 1$ then part (i) of Theorem 1.6 implies that p divides b. ∎

The above result says that if p is a prime and it divides the product of two integers, it must divide one or the other of the integers (or both). Note that this is not necessarily true of divisors that are not prime. For example, $4|12$, which can be written as 6×2. So $4|(6 \times 2)$, but $4 \nmid 6$ and $4 \nmid 2$.

The next result shows that if a prime divides the product of any finite number of integers, it must divide one of them. The reader will be asked to prove this corollary in the exercises.

Corollary 1.8 *If p is a prime and p divides $a_1 \cdots a_r$, then p must divide a_i for some $i = 1, \ldots, r$.*

We now state, and prove, one of the most fundamental results in number theory. This result states that any positive integer can be written as a product of primes, and that with a possible re-ordering of the primes, this product of primes is unique.

Theorem 1.9 *(Unique Factorization)*
   (i) Each positive integer $n \geq 2$ may be written as a product of primes; i.e., n may be written in the form

$$n = p_1 \cdots p_r$$

where each integer $p_i, i = 1, \ldots, r$, is a prime.

(ii) Moreover, this factorization is unique except for the order of the primes; i.e., if $n = q_1 \cdots q_s$ where each q_i is a prime, then $r = s$ and (if necessary) upon re-ordering, $p_i = q_i, i = 1, \ldots, r$.

Proof: For the proof of part (i), we use strong induction. This proof is given in the Appendix, Section 9.1 where strong mathematical induction is discussed.

For the proof of part (ii), we proceed by regular induction (also discussed in the same section of the Appendix). We induct on r, the number of primes in the factorization. For the $r = 1$ case assume that n is a prime and that we also have $n = q_1 \cdots q_s$.

If $s \geq 2$, then the prime n would have at least the three distinct divisors $1, q_1, q_1 q_2$, but this contradicts the fact that n is a prime.

We now use the induction hypothesis that any positive integer greater than 2 that has a factorization into $r - 1$ primes has a unique factorization in the above sense. Assume further that

$$n = p_1 \cdots p_r = q_1 \cdots q_s$$

are two prime factorizations of the positive integer n. The prime p_1 divides n so it must divide one of the primes $q_1, \ldots, q_s$. By re-ordering the primes q_i if necessary, we may assume that p_1 divides q_1. But since both p_1 and q_1 are primes, we must have that $p_1 = q_1$.

Hence we can divide both sides of the above equation by p_1 to obtain

$$p_2 \cdots p_r = q_2 \cdots q_s.$$

The integer on the left side is thus a product of $r - 1$ primes, so by the induction hypothesis, we have that $r - 1 = s - 1$ and hence $r = s$. Upon re-ordering if necessary, we have $p_i = q_i, i = 2, \ldots, r$. Since $p_1 = q_1$, we have that $p_i = q_i, i = 1, \ldots, r$ and the proof is complete. ∎

We note in the statement of the Unique Factorization Theorem that $n \geq 2$. This theorem provides a good reason why 1 is not considered to be a prime. For if 1 was a prime, then for example

$$6 = 2(3) = 1(2)(3)$$

would have two different numbers of primes in its prime factorization, so the factorization would not be unique.

We can, of course, collect like primes in the factorization of an integer n and thus write $n = p_1^{a_1} \cdots p_t^{a_t}$ where each $p_i, i = 1, \ldots, t$ is a prime with $p_i \neq p_j$ if $i \neq j$ and each exponent $a_i \geq 1$. This form is often called the **prime factorization** or **canonical factorization** of the positive integer n.

For example, the prime factorization of the integer $n = 1,000$ is given by $n = 2^3 5^3$ while the prime factorization of the integer $n = 3,500$ can be written as $2^2 5^3 7$.

We should point out here that given a positive integer n, the Unique Factorization Theorem does not give us any idea how to find the primes in the factorization of n. In fact if n is a very large number, this is a formidable problem, even with the use of a fast modern computer. The reader will see in Section 1.6, when we study the RSA cryptographic system, that the security of the system is based upon the difficulty of factoring a very large number.

We close this section by sharing an alternate approach to finding $\gcd(a, b)$ when the prime factorizations of a and b are known.

Theorem 1.10 *Let $a = p_1^{a_1} p_2^{a_2} \cdots p_r^{a_r}$ and $b = p_1^{b_1} p_2^{b_2} \cdots p_r^{b_r}$ be the prime factorizations of a and b, respectively (where perhaps some of the exponents are zero in order to allow a common list of primes to be used). Here, $p_1, p_2, \ldots, p_r$ are distinct primes, $a_1, a_2, \ldots, a_r \geq 0$ and $b_1, b_2, \ldots, b_r \geq 0$. Then*

$$\gcd(a, b) = p_1^{min(a_1, b_1)} p_2^{min(a_2, b_2)} \cdots p_r^{min(a_r, b_r)}$$

where $min(x, y)$ is the smaller of the two values x and y.

Using the above theorem, we can calculate $\gcd(350, 450)$ rather quickly once we do the work of finding the prime factorizations of 350 and 450. Notice that $350 = 2^1 5^2 7^1$ and $450 = 2^1 3^2 5^2$. In order to apply Theorem 1.10, we include each of the primes $2, 3, 5,$ and 7 in both of our prime factorizations. So we rewrite the prime factorizations as $350 = 2^1 3^0 5^2 7^1$ and $450 = 2^1 3^2 5^2 7^0$.

We are now in a position to find the desired gcd. Theorem 1.10 tells us that $\gcd(350, 450) = 2^1 3^0 5^2 7^0$ or $\gcd(350, 450) = 50$.

1.2 Exercises

1. Using the Sieve of Eratosthenes, list all of the primes less than 100.

2. Is 16,667,565 a prime number? What about 1,024,000?

3. Find the canonical factorization of 384.

4. Find the canonical factorization of 3072.

5. Find the canonical factorization of 6250.

6. Find $\gcd(2^3 3^4 5^{10}, 3^4 5^5 7^3)$.

7. Find $\gcd(2^5 3^8 5, 3^3 5^5 7^6)$.

8. Prove that $m = p_1^{a_1} \cdots p_r^{a_r}$ is a square if and only if each a_i is even. Recall that a positive integer b is a **square** if and only if b can be written as $b = a^2$ for some integer a.

9. Show that in using the Sieve of Eratosthenes to obtain all of the primes up to some positive integer n, we only need to sieve by the primes up to the value $\sqrt{n}$.

10. Prove Corollary 1.8.

11. Show that for any positive integer n there exist n consecutive integers, none of which is a prime. As a hint, start with the number $(n+1)! + 2$. Recall that $m! = m(m-1)\cdots(2)(1)$.

1.3   Congruences

Here we discuss an important notion defined on the integers, which was originally developed by Carl Friedrich Gauss in his book *Disquisitiones Arithmeticae*, published in 1801.

Let $n \geq 2$ be a fixed integer. We define two integers a and b to be **congruent modulo** n if n divides the difference $a - b$. We will denote this by writing $a \equiv b \pmod{n}$. We call the integer n the **modulus** of the congruence.

Thus we see that $a \equiv b \pmod{n}$ if $a - b = nk$ for some integer k. Note that it doesn't matter in which order we work, because if $a - b = nk$, then $b - a = n(-k)$, so that $b \equiv a \pmod{n}$.

Note that in our definition we required the modulus n to be at least 2. The reader may wonder why this was done; why not allow n to be any positive integer? We note that given any two integers a and b, we have $a - b = 1(a - b)$ so that if $n = 1$, every integer would be congruent to every other integer modulo n. Not a very interesting, or useful, notion!

Probably without realizing it, the reader has already encountered congruences in everyday life. For example, our clock system works modulo 12 and the 24-hour military clock works modulo 24. Days of the week are determined modulo 7 because if a given day is Monday, then seven days later we have another Monday. Similarly, except for leap years, our yearly calendars work modulo 365.

We now provide a few examples to illustrate the notion of congruences. First, $7 \equiv 2 \pmod 5$ since $7 - 2 = 5(1)$. Similarly, $27 \equiv 5 \pmod{11}$ since $27 - 5 = 22 = 11(2)$. We note that 27 is also congruent to 5 modulo 2 since $27 - 5 = 22 = 2(11)$.

One has to be a bit more careful with 0 and with negative numbers, but the ideas are the same. For example, is $4 \equiv -21 \pmod 5$? Well, we calculate

$$4 - (-21) = 4 + 21 = 25 = 5(5)$$

so yes, 4 and -21 are congruent to one another modulo 5.

Similarly we note that $6 \equiv 0 \pmod 3$ since $6 - 0 = 6 = 3(2)$, but as the reader should check, 36 is not congruent to 0 mod 7.

We note for any integer a, that $a \equiv a \pmod{n}$ for any modulus n since $a - a = 0 = n(0)$.

There is an alternative way to determine if two integers a and b are congruent modulo m. One can simply divide each integer by the modulus m and check to see if the two remainders are the same. For example, we noted above that $27 \equiv 5 \pmod{11}$. Dividing 27 by 11, we obtain a remainder of 5. Similarly, when dividing 5 by 11, we also obtain the same remainder of 5. Because these two remainders are the same, we can conclude that $27 \equiv 5 \pmod{11}$.

As another, larger illustration, consider the positive integers 235 and 147 with $m = 11$. Dividing 235 by 11, we obtain a remainder of 4; similarly dividing 147 by 11 we also obtain a remainder of 4. So 235 and 147 are congruent modulo 11. As a check we can also calculate $235 - 147 = 88 = 11(8)$ so that $235 \equiv 147 \pmod{11}$.

We now state some properties of congruences that will be useful in our later work. We will prove several of these properties, leaving proofs of the remaining properties for the reader in the exercises.

Lemma 1.11 *Let $n \geq 2$ be a fixed integer. Assume that $a \equiv b \pmod{n}$ and that $c \equiv d \pmod{n}$. Then*

(i) $a + c \equiv b + d \pmod{n}$;
(ii) $a - c \equiv b - d \pmod{n}$;
(iii) $ac \equiv bd \pmod{n}$;
(iv) If m is an integer, then $ma \equiv mb \pmod{n}$;
(v) If d is a divisor of n, then $a \equiv b \pmod{d}$.

Proof: We will prove parts (i), (iii), and (v), leaving proofs of the remaining parts to the exercises. From the assumption of the lemma, we first note that $a - b = nk$ and $c - d = nl$ for some integers k and l. To prove part (i), we calculate

$$(a + c) - (b + d) = a + c - b - d = nk + nl = n(k + l).$$

Since $k + l$ is an integer, we can conclude that

$$a + c \equiv b + d \pmod{n}.$$

For part (iii) we note that

$$ac = (b + nk)(d + nl) = bd + bnl + dnk + n^2kl = bd + n(bl + dk + nkl).$$

Therefore
$$ac - bd = n(bl + dk + nkl)$$

where $bl + dk + nkl$ is an integer. Hence $ac \equiv bd \pmod{n}$.

Finally for part (v) we see that $n = de$ for some integer e. Hence we have that

$$a - b = nk = dek = d(ek)$$

so that $a \equiv b \pmod{d}$ since ek is an integer.　　　■

The alert reader will have noticed in the above properties of congruences that one can add, subtract, and multiply congruences and the new congruence will remain true, with the same modulus. You may wonder, what about division? It turns out in the theory of congruences that sometimes one can divide and sometimes one can't!

This will be explained in more detail shortly, but for now note that $40 \equiv 30$ (mod 5). If we divide both sides of the congruence by 2 and leave the modulus the same, we have that $20 \equiv 15$ (mod 5), so in this case we can indeed divide by 2 and the new congruence remains true.

However, note that $40 \equiv 30$ (mod 5), but upon dividing both sides by 10, we obtain the congruence $4 \equiv 3$ (mod 5), which is, of course, not true.

We will discuss this phenomenon in more detail later, but the reader should not be surprised that when working with congruences, one cannot always divide. Something similar happens with the real numbers: If we have $ac = bc$, we cannot always divide by c to obtain $a = b$ unless we have $c \neq 0$.

When can one divide in a congruence and keep the modulus the same? This question is answered in the following lemma.

Lemma 1.12 *If* $ac \equiv bc$ (mod n) *and* c *and* n *are relatively prime, then* $a \equiv b$ (mod n).

Proof: We have
$$ac - bc = (a - b)c = nk$$
for some integer k. Thus n divides $(a-b)c$. Since c and n are relatively prime, from part (i) of Theorem 1.6, n must divide $a - b$. Thus $a \equiv b$ (mod n). ∎

We note in passing that if we divide both sides of the congruence
$$ac \equiv bc \pmod{n}$$
by d where d is any common divisor of c and n, we have
$$\frac{ac}{d} \equiv \frac{bc}{d} \pmod{\frac{n}{d}}.$$

This result follows from the definition of congruence. Note that here the modulus has changed from n to n/d.

Assume that we have a congruence in which the modulus is n. Given any positive integer a, one can always find an integer r with $0 \leq r < n$ so that
$$a \equiv r \pmod{n}.$$

Such an integer r is called the **least non-negative residue of** a **modulo** n. It is often very convenient in performing congruence calculations to work with the least non-negative residue modulo n.

Given an integer a, how can one find this least non-negative residue modulo n? If a is positive, one method is to simply use long division (the Division Algorithm, Theorem 1.1) to divide the integer a by n. Then the remainder r is the least non-negative residue modulo n. (Recall from the Division Algorithm that the remainder r when dividing a by n has the property that $0 \le r < n$.)

If the integer a is negative, we can always add enough multiples of n until the value becomes in the range $0 \le r < n$.

We now consider a few examples.

Example 1.13 *The least non-negative residue of 27 modulo 6 is 3 since* $27 = 6(4) + 3$ *and* $0 \le 3 < 6$. *Similarly the least non-negative residue of 27 modulo 9 is 0 since* $27 = 9(3) + 0$.

Example 1.14 *If* $a = -22$, *then the least non-negative residue of a modulo 8 is 2 since* $-22 + 24 \equiv 2 \pmod 8$ *and* $0 \le 2 < 8$. *One can check that this is correct by noting that* $-22 \equiv 2 \pmod 8$, *since*

$$-22 - 2 = -24 = 8(-3).$$

1.3 Exercises

1. Which of the following congruences hold?

    - $17 \equiv 5 \pmod 9$
    - $33 \equiv 0 \pmod{11}$
    - $55 \equiv -9 \pmod{16}$
    - $283 \equiv 177 \pmod 5$
    - $283 \equiv 177 \pmod 2$
    - $220 \equiv 14 \pmod 6$
    - $34 \equiv -12 \pmod{23}$
    - $17 \equiv -35 \pmod 9$
    - $3m + 1 \equiv 2m + 1 \pmod m$ for any integer $m \ge 2$
    - $3m + 3 \equiv 4m + 3 \pmod m$ for any integer $m \ge 2$

2. For each of the following congruences, fill in the blank with the least non-negative residue:

    - $17 \equiv$ _____ $\pmod 9$
    - $21 \equiv$ _____ $\pmod 7$
    - $0 \equiv$ _____ $\pmod 9$
    - $-24 \equiv$ _____ $\pmod 6$
    - $334 \equiv$ _____ $\pmod{55}$
    - $220 \equiv$ _____ $\pmod 6$
    - $-221 \equiv$ _____ $\pmod{33}$

- $-225 \equiv$ _____ $(\text{mod } 8)$
- $7 + 257 \equiv$ _____ $(\text{mod } 12)$
- $-22 - 333 \equiv$ _____ $(\text{mod } 8)$

3. Prove part (ii) of Lemma 1.11.

4. Prove part (iv) of Lemma 1.11.

5. Fix an integer $n \geq 2$. Show that the relation on the integers defined by saying that two integers a and b are related if $a' \equiv b \pmod{n}$ is an equivalence relation. See the Appendix, Section 3 for a discussion of equivalence relations.

6. For any positive integer n, find the least non-negative residue of 6^n modulo 10.

7. Using congruences, prove that $8^n - 1$ is divisible by 7 for all positive integers n.

8. Using congruences, prove that $5|(18^n - 3^n)$ for all positive integers n.

9. Show that if $a \equiv b \pmod{n}$, then $a^e \equiv b^e \pmod{n}$ for any positive integer e.

1.4   Solving congruences

In this section we will discuss when linear congruences; i.e., congruences of the form $ax \equiv b \pmod{n}$, have solutions x, and if such a congruence has one or more solutions, we will discuss how to find all of the solutions. This will be of importance later in several parts of this text.

Consider the congruence $ax \equiv b \pmod{n}$ where a, b, and n are fixed integers. By a **solution** of the congruence we mean an integer x that satisfies the congruence. We first determine when such a congruence has a solution. To do so, we calculate $\gcd(a, n)$ and call this value d. The existence of a solution to our congruence then hinges on an important relationship between b and d.

Theorem 1.15 *(i) A linear congruence $ax \equiv b \pmod{n}$ has solutions if and only if the greatest common divisor d of a and n divides b.*

(ii) If d divides b, then there are d solutions modulo n (i.e., each of the d solutions x has the property that $0 \leq x < n$ and the solutions are congruent modulo $\frac{n}{d}$).

Proof: If $ax \equiv b \pmod{n}$, then $ax - b = nk$ for some integer k, and hence $ax - nk = b$ must be divisible by d since d divides the left-hand side. (Recall that $d|a$ and $d|n$ so that d divides the quantity $ax - nk$.) Thus if d fails to divide b, then the congruence $ax \equiv b \pmod{n}$ has no solution and we are done.

Assume now that d divides b so that $b = de$ for some integer e. Since d is the greatest common divisor of a and b, we can write d as $d = ak + nl$ for integers k and l. Hence we know $b = ake + nle$. This can be re-written as a congruence by observing that $a(ke) - b = n(le)$, so that

$$a(ke) \equiv b \pmod{n}.$$

This shows that the integer ke is a solution of the original congruence $ax \equiv b \pmod{n}$.

To prove part (ii) of the theorem, let c be a solution of the congruence $ax \equiv b \pmod{n}$. Then $ac = b + nr$ for some integer r. Since d divides b, and since $d = \gcd(a, n)$, we may divide this equation by d to obtain

$$(a/d)c = b/d + (n/d)r.$$

Note that $a/d, b/d$, and n/d are integers. Thus we have

$$(a/d)c \equiv b/d \pmod{n/d}.$$

Hence every solution of the original congruence is also a solution of this congruence. Moreover, it can be checked that every solution of this congruence is also a solution of the congruence $ax \equiv b \pmod{n}$.

If c is any solution of the original congruence, then

$$c, c + (n/d), c + 2(n/d), \dots, c + (d-1)(n/d)$$

are all distinct solutions modulo n. This gives d solutions of the original congruence. ∎

We now provide several examples to illustrate the above theorem.

Example 1.16 *Consider the congruence $4x \equiv 6 \pmod{8}$. We note that $\gcd(4, 8) = 4$ and that 4 does not divide 6. So by part (i) of Theorem 1.6, the congruence does not have any solutions. The reader should also verify that this congruence does not have any solutions by substituting each of the values $0 \leq x < 8$ into the congruence and noticing that none of them solve the congruence.*

Example 1.17 *The congruence $3x \equiv 5 \pmod{7}$ has exactly one solution in the range $0 \leq x < 7$ since $\gcd(3, 7) = 1$ and 1 clearly divides 5. The value $x = 4$ satisfies the congruence since $3(4) - 5 = 1(7)$.*

Example 1.18 *Consider the congruence $4x \equiv 8 \pmod{12}$. We note that $d = \gcd(4, 12) = 4$ and 4 divides 8. So by Theorem 1.15 there should be 4 different solutions modulo 12; i.e., 4 different solutions in the range $0 \leq x < 12$.*

We note that $n/d = 12/4 = 3$. We also note that $c = 2$ is a solution. So, the four different solutions are given by

$$c, c + (n/d), c + 2(n/d), c + 3(n/d).$$

In other words, the four solutions of the original congruence are given by

$$2 + 0(3), 2 + 1(3), 2 + 2(3), \text{ and } 2 + 3(3)$$

or

$$2, 5, 8, \text{ and } 11.$$

We next consider the possibility of solving simultaneous linear congruences. The result appears to have been first published by the Chinese mathematician Sun Tzu, who lived sometime between the third and fifth centuries A.D.

Theorem 1.19 *(Chinese Remainder Theorem)* *Let $m \geq 2$ and $n \geq 2$ be integers that are relatively prime. Let a and b be integers. Then there is a simultaneous solution to the pair of congruences*

$$x \equiv a \pmod{m}$$

$$x \equiv b \pmod{n}.$$

Moreover, this solution is unique modulo mn, i.e., there is only one solution x with $0 \leq x < mn$.

Proof: Since m and n are relatively prime, there are integers r and s so that $mr + ns = 1$. We can check that $c = bmr + ans$ is a simultaneous solution to the pair of congruences. We know that $c \equiv ans \pmod{m}$ and that $ns \equiv 1 \pmod{m}$. Thus $c \equiv a(1) \pmod{m}$. The proof that c is also a solution of the second congruence is similar and is left to the reader.

We now show that our solution is unique. To this end, assume that c and d are both solutions. Then $c \equiv a \pmod{m}$ and $d \equiv a \pmod{m}$. Then $c - d \equiv 0 \pmod{m}$ and similarly $c - d \equiv 0 \pmod{n}$. Thus $c - d$ is divisible by both m and n, and since m and n are relatively prime, $c - d$ is divisible by the product mn. Hence $c \equiv d \pmod{mn}$. ∎

Example 1.20 *As a small illustration, consider the simultaneous pair of congruences*

$$x \equiv 3 \pmod{4}$$
$$x \equiv 4 \pmod{5}.$$

*Starting with the larger modulus we have $x - 4 = 5k$ so that $x = 5k + 4$
for some integer k. Substituting this into the other congruence, we obtain
$5k + 4 \equiv 3 \pmod 4$, and after reducing the coefficents modulo 4 we have
$k \equiv -1 \equiv 3 \pmod 4$. Thus $k = 3$ is the least non-negative residue modulo 4.
Then $x = 5(3) + 4 = 19$. Note that $0 \leq 19 < 4(5) = 20$ so our solution x lies
in the correct range.*

*We now check that we have done our arithmetic correctly by calculating
that $19 - 3 = 16 = 4(4)$ so $x \equiv 3 \pmod 4$. Similarly we can check that $19 \equiv 4$
$\pmod 5$ so $x = 19$ satisfies both congruences.*

Note that in finding the above solution, analogous to solving a pair of
simultaneous linear equations in the variables x and y, we first solved for
x in one of the congruences (to save arithmetic, one should do this in the
congruence with the larger modulus). We then substituted this expression for
x in the other congruence and solved this second congruence.

Example 1.21 *For a slightly larger illustration, consider the simultaneous
pair of congruences*

$$x \equiv 5 \pmod 7$$
$$x \equiv 13 \pmod{19}.$$

*From the congruence with the larger modulus we have $x - 13 = 19k$ for
some integer k. Then $x = 19k + 13$. Substituting this into the other congruence
we obtain $19k + 13 \equiv 5 \pmod 7$. Hence after some arithmetic modulo 7, we
have that $5k \equiv 6 \pmod 7$. Thus $k = 4$ is a solution to this congruence. Hence
$x = 19k + 13 = 19(4) + 13 = 89$ is the simultaneous solution to our system of
linear congruences.*

*We note that $0 \leq 89 < 7(19) = 133$, so that our solution x indeed lies
in the correct range. To check our work, we quickly find that $89 - 5 = 84$ is
divisible by 7 so $x = 89$ satisfies the first congruence. Similarly $89 - 13 = 76$
is divisible by 19 so x also satisfies the second congruence.*

We should point out that if m and n are not relatively prime, i.e., if
$\gcd(m, n) \neq 1$, a pair of simultaneous congruences might not have a solution.
For example, consider the pair of simultaneous congruences

$$x \equiv 1 \pmod 2$$
$$x \equiv 2 \pmod 4.$$

We note that if x were a simultaneous solution, then x would have the properties that $x = 2k + 1$ for some integer k and $x = 4l + 2$ for some integer l.
But $2k + 1$ can never equal $4l + 2$ for any integers k and l. This is because
for any integer k, the integer $2k + 1$ is always odd, while the integer $4l + 2$ is
always even for any integer l.

The reader may wonder about the problem of solving three or more simultaneous congruences. This can be done as long as the moduli are all relatively

prime. We will illustrate how to solve a set of three simultaneous congruences. The technique demonstrated here can be generalized to a system that contains numerous linear congruences of the form $x \equiv a \pmod{m}$.

Consider the simultaneous congruences

$$x \equiv 3 \pmod{5}$$
$$x \equiv 4 \pmod{6}$$
$$x \equiv 5 \pmod{7}.$$

Starting with the largest modulus (to save calculation and to be able to work with smaller moduli later), we may assume that $x = 5 + 7k$ for some integer k. Then, substituting this into the congruence with the next largest modulus (6 in our case), we obtain

$$5 + 7k \equiv 4 \pmod{6}$$

, so that after simplification we have the congruence $k \equiv 5 \pmod{6}$. Hence $k = 5$. Then $x = 5 + 7(5) = 40$ is the solution to the last pair of congruences. Recall that the solution for this last pair of congruences should be in the range $0 \le x < 6(7) = 42$.

We must still take into account the first congruence, so we assume that $x = 40 + 42n$ for some integer n. Substituting this into the first congruence, we obtain

$$40 + 42n \equiv 3 \pmod{5}$$

After simplification, this is equivalent to the congruence $2n \equiv 3 \pmod{5}$ and hence $n = 4$.

Finally,

$$x = 40 + 42n = 40 + 42(4) = 208.$$

Note that this value is in the correct range: $0 \le x < 5(6)(7) = 210$. As with any computational problem of this type, we should check that the solution is indeed correct. To this end, note that $208 - 3 = 205$ is divisible by 5, $208 - 4 = 204$ is divisible by 6, and finally, $208 - 5 = 203$ is divisible by 7. Hence x is the unique solution to the system of three congruences.

1.4 Exercises

1. Determine if the following congruences have solutions. If they do, determine the number of solutions smaller than the modulus and then find all solutions less than the modulus of the given congruence.

   (a) $5x \equiv 6 \pmod{12}$

   (b) $5x \equiv 12 \pmod{20}$

   (c) $4x \equiv 12 \pmod{20}$

   (d) $12x \equiv 14 \pmod{16}$

   (e) $12x \equiv 24 \pmod{144}$

2. Find a common solution to the pair of congruences

$$x \equiv 4 \pmod{7}$$
$$x \equiv 5 \pmod{11}.$$

3. Find a common solution to the pair of congruences

$$x \equiv 5 \pmod{11}$$
$$x \equiv 6 \pmod{17}.$$

4. Find a common solution to the pair of congruences

$$x \equiv 2 \pmod{6}$$
$$2x \equiv 1 \pmod{7}.$$

Hint: Consider multiplying both sides of the second congruence by 4 and then proceeding. Why do we multiply by the value 4?

5. Solve the system of congruences

$$x \equiv 5 \pmod{8}$$
$$x \equiv 4 \pmod{9}$$
$$x \equiv 5 \pmod{11}.$$

6. Find the smallest positive integer whose remainder when divided by 11 is 8, which has last digit 4, and is divisible by 27.

1.5   Theorems of Fermat and Euler

We now develop a method to evaluate powers in congruences. For example, consider the prime $p = 5$. How does one efficiently determine the least non-negative residue of 3^4 modulo 5? Or worse yet, how do we handle something like 3^{100} modulo 5? In the first case, one could calculate $3^4 = 81$ so that upon division by 5, we obtain the remainder 1. Thus $3^4 \equiv 1 \pmod{5}$. But how do we determine the least non-negative residue of 3^{100} modulo 5? Surely there is a better and faster way to obtain this least non-negative residue than to multiply out the value 3^{100}. Fortunately, there is indeed a much faster method, first discovered by the French lawyer, government official, and amateur mathematician Pierre de Fermat (1601 – 1665).

Theorem 1.22 *(Fermat) If p is a prime and $gcd(a, p) = 1$, then*

$$a^{p-1} \equiv 1 \pmod{p}.$$

Proof: Consider the product $1(2)(3)\cdots(p-1)$. Also consider the product of each of these elements multiplied by the number a, i.e., consider the product

$$(1a)(2a)(3a)\cdots((p-1)a).$$

These multiples of a must be distinct modulo the prime p, for otherwise, if

$$ia \equiv ja \pmod{p},$$

then since a is relatively prime to p, we can divide by a to obtain $i \equiv j \pmod{p}$ thanks to Lemma 1.12. But this is a contradiction, since $1 \le i, j \le p-1$.

Thus this new set of integers $\{1a, 2a, \ldots, ((p-1)a)\}$ must be the same modulo p, except for the order, as the set $\{1, 2, \ldots, p-1\}$. Hence, we must have that

$$(1a)(2a)(3a)\cdots((p-1)a) \equiv 1(2)(3)\cdots(p-1) \pmod{p}.$$

Notice now that we have the common factors $1, 2, \ldots, p-1$ on both sides of the congruence. Each of the values is greater than zero and less than the prime p. Hence, each value is relatively prime to p. We can thus divide each of those values from both sides of the congruence (again thanks to Lemma 1.12) to obtain

$$\underbrace{a(a)\cdots(a)}_{p-1 \text{ times}} \equiv \underbrace{1(1)\cdots(1)}_{p-1 \text{ times}} \pmod{p}.$$

Hence, we see that $a^{p-1} \equiv 1 \pmod{p}$ and the proof is complete. $\blacksquare$

We note that Fermat's Theorem implies, for any integer a and any prime p, that $a^p \equiv a \pmod{p}$. If a is not divisible by p, then Fermat's Theorem implies that $a^{p-1} \equiv 1 \pmod{p}$, and thus by multiplying both sides by p, we have $a^p \equiv a \pmod{p}$. If p divides a, then $a^p \equiv 0 \pmod{p}$ and $a \equiv 0 \pmod{p}$, so $a^p \equiv a \pmod{p}$.

Example 1.23 *By way of illustration, using Fermat's Theorem, we know that $3^4 \equiv 1 \pmod 5$ without having to do any work because $\gcd(3,5) = 1$. Similarly since 17 is a prime, Fermat's Theorem tells us that $3^{16} \equiv 1 \pmod{17}$, also without any work (since $\gcd(3,17) = 1$). This is significant since $3^{16} = 43,046,721$ and dividing this number by 17 to find the remainder would be extremely time-consuming.*

Returning to our earlier example, how do we handle a situation like 3^{100} modulo 5? We make use of the fact that from Fermat's Theorem we have $3^4 \equiv 1 \pmod 5$. Note that $100 = 4(25)$, so that we may calculate

$$3^{100} \equiv 3^{4(25)} \equiv (3^4)^{25} \equiv 1^{25} \equiv 1 \pmod 5$$

and we are done!

What if our exponent is not divisible by 4? Consider $103 = 4(25)+3$. Then

$$3^{103} \equiv 3^{4(25)+3} \equiv 3^{4(25)}3^3 \equiv (3^4)^{25}3^3 \equiv 1^{25}3^3 \equiv 3^3 \equiv 2 \pmod 5$$

since $27 \equiv 2 \pmod 5$. *Thus,* $3^{103} \equiv 2 \pmod 5$. *This means that if we divide* 3^{103} *by 5, the remainder is 2.*

A natural question to ask at this point is the following: What if our modulus is not a prime? For example, how do we handle something like 5^{29} (mod 12)? What is the least non-negative residue of 5^{29} modulo 12?

To answer such a question, we now consider one of the most important functions in elementary number theory. This function is called **Euler's** ϕ **function**, named in honor of Leonard Euler $(1707 - 1783)$. Given a positive integer n we define the **Euler function** $\phi(n)$ to be the number of positive integers less than or equal to n that are relatively prime to n. Thus $\phi(n)$ counts the number of positive integers k with $1 \le k \le n$ which have the property that $\gcd(k, n) = 1$.

For example, we see that $\phi(5) = 4$ since the positive integers $1, 2, 3$, and 4 are each relatively prime to 5. Similarly, $\phi(8) = 4$ s ince the positive integers $1, 3, 5$, and 7 are each relatively prime to 8. The reader should check that we also have $\phi(12) = 4$.

Given a positive integer n, how does one determine the value $\phi(n)$? For small values of n, one could, of course, list all of the positive integers less than or equal to n and simply count the number of these integers that are relatively prime to n. This is feasible to do for small values of n like we did above, but is there a faster way to determine the value $\phi(n)$ without checking each positive integer less than n to see if it is relatively prime to n? Thankfully, the answer is yes!

The following three theorems tell us how to calculate Euler's function $\phi(n)$ for any positive integer n.

Theorem 1.24 *If p is a prime, then $\phi(p) = p - 1$.*

Proof: Recall from the definition that $\phi(p)$ counts the number of positive integers up to p which are relatively prime to p. Clearly each value between 1 and $p - 1$ is relatively prime to p. Moreover, $\gcd(p, p) = p$, which is not 1, so we do not count p. Thus $\phi(p) = p - 1$, which completes the proof. ∎

This result can be extended to powers of primes as follows:

Theorem 1.25 *If p is a prime and a is a positive integer, then*

$$\phi(p^a) = p^a - p^{a-1}.$$

Proof: There are p^a positive integers k with $1 \le k \le p^a$. Of these, $1p, 2p, \ldots, p^{a-1}p$ are each divisible by the prime p and hence not relatively prime to p^a. There are p^{a-1} values in this list. The remaining values are each relatively prime to p^a. Hence $\phi(p^a) = p^a - p^{a-1}$. ∎

Remark 1.26 *Note that if $a = 1$, Theorem 1.25 reduces to Theorem 1.24.*

Theorems 1.24 and 1.25 are easy to use for primes and prime powers, but how does one calculate the value $\phi(n)$ when n is not a prime power? For example, how does one calculate $\phi(12)$? We know by checking the positive integers up to 12 that there are exactly four values of k with $\gcd(k, 12) = 1$; namely, the values $1, 5, 7$, and 11. Thus, $\phi(12) = 4$. Is there a faster way to obtain the fact that $\phi(12) = 4$ without checking each value up to 12? Again, the answer is yes, and the next result tells us how to do this.

Theorem 1.27 *If a and b are relatively prime positive integers, then $\phi(ab) = \phi(a)\phi(b)$.*

Proof: Instead of giving a formal proof of this result, we illustrate the method of proof with an example. The main idea behind the proof involves the use of the Chinese Remainder Theorem studied in Theorem 1.19.

Assume that we wish to calculate $\phi(12)$. The positive integers k with $1 \le k \le 12$, which have the property that k and 12 are relatively prime, are the integers $1, 5, 7$, and 11. Thus $\phi(12) = 4$ since $\phi(12)$ counts the number of such positive integers.

Clearly $12 = 3(4)$, and 3 and 4 are relatively prime. We now consider all ordered pairs (a, b) with $1 \le a \le 3$ and a relatively prime to 3. Thus $a = 1$ or 2. Similarly, let $1 \le b \le 4$ with b and 4 relatively prime. That means $b = 1$ or 3. We now solve, using the Chinese Remainder Theorem, the pair of simultaneous congruences

$$x \equiv a \pmod{m}$$

$$x \equiv b \pmod{n}.$$

We first consider the pair of congruences

$$x \equiv 1 \pmod 3$$

$$x \equiv 1 \pmod 4$$

whose unique solution x is given by $x = 1$ modulo 12.

Similarly, the pair of congruences

$$x \equiv 2 \pmod 3$$

$$x \equiv 1 \pmod 4$$

has the unique solution $x = 5$ modulo 12.

The pair of congruences

$$x \equiv 1 \pmod 3$$

$$x \equiv 3 \pmod 4$$

has the unique solution $x = 7$ modulo 12.

Finally, the pair of congruences

$$x \equiv 2 \quad (\text{mod } 3)$$

$$x \equiv 3 \quad (\text{mod } 4)$$

has the unique solution $x = 11$ modulo 12.

Notice that each of the four pairs (a, b) gives, by the Chinese Remainder Theorem, a unique integer relatively prime to 12, i.e., in our example we obtain the four values $1, 5, 7, 11$.

Similarly, each of the values $1, 5, 7, 11$, which are relatively prime to 12, yields a unique pair (a, b) with a relatively prime to 3, and b relatively prime to 4. Hence we have the same number of pairs (a, b) as there are values of k with $1 \leq k \leq 12$ that are relatively prime to 12.

Thus we have shown that $\phi(12) = \phi(3)\phi(4)$, as desired.

As long as a and b are relatively prime, the same argument will work to show that $\phi(ab) = \phi(a)\phi(b)$. ∎

We now illustrate these important properties by providing several examples related to the determination of the values of Euler's function $\phi(n)$.

Example 1.28 *First we note from Theorem 1.24 that $\phi(3) = 2$ and $\phi(17) = 16$ since 3 and 17 are both primes. To illustrate Theorem 1.25, note that in our earlier example we found that $\phi(8) = 4$. Using Theorem 1.25, we see that since 2 is a prime, $8 = 2^3$ and hence*

$$\phi(8) = \phi(2^3) = 2^3 - 2^2 = 8 - 4 = 4$$

as noted earlier. Similarly

$$\phi(81) = \phi(3^4) = 3^4 - 3^3 = 81 - 27 = 54.$$

Note that by using Theorem 1.25, we are able to determine the value $\phi(81) = 54$ without listing the 54 values of k with $1 \leq k \leq 81$ with the property that $gcd(k, 81) = 1$.

Example 1.29 *For cases where n is not a prime power, one proceeds as follows. First one determines the prime factorization of n. Then we apply Theorem 1.26 to split the calculation into smaller pieces, and then we apply Theorem 1.25 to each prime power. Recall that in the prime factorization of any positive integer, the prime powers are each relatively prime to each other, so we can apply Theorem 1.25 to each individual prime power.*

To illustrate, we determine $\phi(12)$. First we obtain the prime factorization of $12 = 2^2(3)$. We now apply Theorem 1.26 to obtain

$$\phi(12) = \phi(2^2(3)) = \phi(2^2)\phi(3).$$

By Theorem 1.25 we then have

$$\phi(12) = \phi(2^2)\phi(3) = (2^2 - 2^1)(3^1 - 3^0) = (4 - 2)(3 - 1) = 2(2) = 4,$$

which agrees with our earlier determination of $\phi(12)$.

Example 1.30 *Similarly, since* $144 = 16(9) = 2^4(3^2)$ *is the prime factorization of 144, we have*

$$
\begin{aligned}
\phi(144) &= \phi(16(9)) \\
&= \phi(2^4(3^2)) \\
&= \phi(2^4)\phi(3^2) \\
&= (2^4 - 2^3)(3^2 - 3^1) \\
&= 48.
\end{aligned}
$$

Equipped with Euler's function $\phi(n)$, we can now extend Fermat's Theorem (Theorem 1.22) to a much more general setting.

Theorem 1.31 *(Euler) If $n \geq 2$ is an integer and $gcd(a, n) = 1$, then $a^{\phi(n)} \equiv 1 \pmod{n}$.*

Proof: The proof of Euler's Theorem is similar to the proof of Fermat's Theorem. Let

$$a_1, a_2, \ldots, a_{\phi(n)}$$

be the set of elements between 1 and n that are relatively prime to n. (Recall that there are exactly $\phi(n)$ such elements, where $\phi(n)$ denotes Euler's function.)

Now consider this same set of elements, but multiply each of them by the value a, which we know to be relatively prime to n. If two of these new elements were congruent modulo n, i.e., if say $aa_i \equiv aa_j \pmod{n}$ for some $1 \leq i < j \leq \phi(n)$, then we would have $a_i \equiv a_j \pmod{n}$ (since $gcd(a, n) = 1$), and this is a contradiction.

Hence it must be the case that

$$a_1 a_2 \cdots a_{\phi(n)} \equiv (aa_1)(aa_2) \cdots (aa_{\phi(n)}) \pmod{n}.$$

Since each of the values $a_1, a_2, \ldots, a_{\phi(n)}$ is relatively prime to the modulus n, they can all be cancelled on both sides of the congruence. Hence on the left side of the congruence we are left with the product of $\phi(n)$ 1's, which is of course congruent to 1 modulo n. On the right side we are left with the product of a with itself $\phi(n)$ times, i.e., $a^{\phi(n)}$. Hence $a^{\phi(n)} \equiv 1 \pmod{n}$ and the proof is complete. ∎

We note that Euler's Theorem is indeed a generalization of Fermat's Theorem because in the case when $n = p$ is a prime, we know from Theorem 1.24 that $\phi(n) = \phi(p) = p - 1$.

We now give some examples to illustrate Theorem 1.31.

Example 1.32 *What is the least non-negative residue of 5^{14} modulo 12? We first note that 5 and 12 are relatively prime, so Euler's Theorem applies. We note from our earlier calculation that $\phi(12) = 4$, so by Euler's Theorem $5^4 \equiv 1 \pmod{12}$. We also note from the Division Algorithm that $14 = 4(3) + 2$, so that*

$$5^{14} \equiv 5^{4(3)+2} \equiv (5^4)^3 5^2 \equiv 1^3 5^2 \equiv 25 \equiv 1 \pmod{12}.$$

Example 1.33 *As a larger example, what is the least non-negative residue of* 2^{314} *modulo 21? Clearly,* $gcd(2, 21) = 1$, *so Euler's Theorem applies. We note that*

$$\phi(21) = \phi(3(7)) = \phi(3)\phi(7) = (3-1)(7-1) = 2(6) = 12.$$

Thus by Euler's Theorem, $2^{12} \equiv 1 \pmod{21}$. *By the Division Algorithm* $314 = 12(26) + 2$. *We thus have*

$$2^{314} \equiv 2^{12(26)+2} \equiv (2^{12})^{26} 2^2 \equiv 1^{26} 2^2 \equiv 4 \pmod{21}.$$

As we close this section, we consider a special set of values known as the *integers modulo n*, often denoted $\mathbb{Z}_n$. In essence,

$$\mathbb{Z}_n = \{0, 1, 2, \ldots, n-1\}$$

where addition of two elements and multiplication of two elements within the set is performed modulo n. So, for example, in $\mathbb{Z}_8$, $5 + 6$ is identified with the number 3 since $5 + 6 = 11 \equiv 3 \pmod{8}$. Similarly, in $\mathbb{Z}_8$, 3×7 is identified with the number 5 since $3 \times 7 = 21 \equiv 5 \pmod{8}$.

We now study the question of when an element a in $\mathbb{Z}_n$ has a multiplicative inverse. An element $a \in \mathbb{Z}_n$ is **invertible** if there is an element $b \in \mathbb{Z}_n$ so that $ab \equiv 1 \pmod{n}$. The element b is the **inverse** of a modulo n. Invertible elements are also often called **units**.

For example, $3(5) \equiv 1 \pmod{7}$ so that modulo 7, 3 is invertible and its inverse is 5 modulo 7. Similarly, 5 is also invertible modulo 7 and its inverse is 3 modulo 7. On the other hand, there is no element b so that $2b \equiv 1 \pmod{6}$, so 2 does not have an inverse modulo 6.

When does an element a in $\mathbb{Z}_n$ have an inverse modulo n? The following result tells us when an inverse exists.

Theorem 1.34 *Let n be a positive integer and let $a \in \mathbb{Z}_n$. Then a has an inverse modulo n if and only if a and n are relatively prime, i.e., if and only if $gcd(a, n) = 1$.*

Proof: If a and n are relatively prime, then there are integers r and s so that $ar + ns = 1$. If we reduce the coefficients modulo n, we obtain $ar \equiv 1 \pmod{n}$ since $n \equiv 0 \pmod{n}$. Hence r is the inverse of a modulo n. Thus, if a and n are relatively prime, then a has an inverse modulo n.

Now assume that b is the inverse of the element a modulo n so that $ab \equiv 1 \pmod{n}$. Let d be the greatest common divisor of a and n. We need to show that $d = 1$. Assume that $d > 1$ and let p be a prime dividing d so that the prime p must also divide a and n.

Since $ab \equiv 1 \pmod{n}$, we know $ab - nk = 1$ for some integer k. Since $p|a$ and $p|n$, we know $p|(ab - nk)$. Thus, $p|1$. This is a contradiction since the prime p must be at least 2. Hence the only common divisor of a and n is 1 and thus a and n are relatively prime. ∎

Example 1.35 *As an illustration of how to compute the inverse modulo n, let $n = 29$ and let $a = 3$. By the Euclidean Algorithm we have*

$$
\begin{aligned}
29 &= 9(3) + 2 \\
3 &= 2(1) + 1 \\
2 &= 2(1) + 0
\end{aligned}
$$

so that $\gcd(29, 3) = 1$.

Using the calculations from the Euclidean Algorithm in reverse, we have that

$$1 = 3 - 2 = 3 - (29 - 3(9)) = 10(3) - 1(29).$$

Hence reducing modulo 29 we obtain the congruence $3(10) \equiv 1 \pmod{29}$ so that the inverse of 3 modulo 29 is 10.

In Section 1.6 we will find that the calculation of inverses is of great importance in the RSA cryptographic system for the secure transmission of information.

Before closing this section, we illustrate an important piece of motivation for our later study of "groups."

Theorem 1.36 *If a and b are invertible modulo n, then the product ab is also invertible modulo n.*

Proof: If a and b are invertible, then from Theorem 1.33 we have that a and n are relatively prime as are b and n. In order to show that the product element ab is invertible modulo n, we must show that ab and n are relatively prime. Suppose that ab and n are not relatively prime. Then there is a prime p that divides both ab and n.

Hence from Theorem 1.7, the prime p must divide a or b and p also divides n. If p divides a and p divides n, then a and n are not relatively prime, contradicting the fact that a is invertible modulo n. Similarly if p divides both b and n, then b is not invertible modulo n, also a contradiction. ∎

The above theorem proves that if a and b are both invertible modulo n, then their product is also invertible modulo n. Such a property, often referred to as a "closure" property, will prove extremely important in later chapters of this book. In particular, we will see the closure property in Chapters 2, 3, and 4 when we consider the algebraic structures known as groups, rings, and fields.

Before closing this section, we refer to several additional books that discuss various aspects of elementary number theory. These textbooks include [1], [11], and [28].

1.5 Exercises

1. In each of the following, fill in the blank with the least non-negative residue.

(a) $2^7 \equiv$ _____ (mod 5)

(b) $3^{14} \equiv$ _____ (mod 7)

(c) $16^{22} \equiv$ _____ (mod 19)

2. In each of the following, fill in the blank with the least non-negative residue.

(a) $3^{21} \equiv$ _____ (mod 14)

(b) $3^{21} \equiv$ _____ (mod 33)

(c) $4^{110} \equiv$ _____ (mod 28)

3. Find $\phi(10), \phi(24), \phi(324)$, and $\phi(1024)$ where ϕ denotes Euler's function.

4. Does $a = 4$ have an inverse modulo 17? If so, find it.

5. Does $a = 6$ have an inverse modulo 28? If so, find it.

6. Does $a = 17$ have an inverse modulo 144? If so, find it.

7. List all of the invertible elements in $\mathbb{Z}_{28}$.

8. List all of the invertible elements in $\mathbb{Z}_{36}$.

9. Assume that a positive integer n is divisible by an odd number which is greater than 1. Show that $\phi(n)$ must be even.

10. Explain why every value $1 \leq a < 4999$ is invertible modulo 4999.

11. If p is a prime, show that every non-zero element of $\mathbb{Z}_p$ is invertible.

12. Show that if a has an inverse modulo n, then that inverse must be unique in $\mathbb{Z}_n$.

1.6   RSA cryptosystem

In today's modern communications, one often needs to send information from one place to another. In addition, this must be done securely so that if unauthorized people are listening in, they are not able to decipher the scrambled data to get the original message. In the system described below, Euler's Theorem (Theorem 1.30) plays a pivotal role. This kind of a system is indeed in use today for various types of electronic transfers (like securely using your credit card when making online purchases).

Before discussing the famous RSA cryptographic system, we illustrate a simple cryptographic system that uses properties from elementary number theory.

We choose positive integers n and s such that $\gcd(s, \phi(n)) = 1$ and with the property that no prime dividing n is smaller than the largest possible message "word." This condition is needed in order to be able to decipher our scrambled messages and always be sure to obtain the original messages by using Euler's Theorem.

Since $\gcd(s, \phi(n)) = 1$, there is an integer t with

$$st \equiv 1 \pmod{\phi(n)}.$$

The integer t is the inverse of s modulo $\phi(n)$.

Hence from our study of number theory, since $\gcd(s, \phi(n)) = 1$ and $st + \phi(n)u = 1$ for some integers t and u, we know that

$$(m^s)^t \equiv m^{1-\phi(n)u} \equiv m(m^{\phi(n)})^{-u} \equiv m \pmod{n}.$$

In the last step of this calculation we have made use of Euler's Theorem, which shows that if m and n are relatively prime, $m^{\phi(n)} \equiv 1 \pmod{n}$.

A message m is enciphered as $m^s \equiv v \pmod{n}$. The value v is called the **ciphertext**. This is the scrambled version of the original message m. A received ciphertext v is deciphered as $v^t \equiv m \pmod{n}$.

Example 1.37 *Let $n = 101$ and let $s = 3$. Since 101 is a prime, $\phi(n) = 100$, and after some calculation, we find that the congruence $st \equiv 1 \pmod{\phi(n)}$ has a solution $t = 67$.*

Assume that we replace the letters $A, B, \ldots, Z$ by $01, 02, \ldots, 26$. Then AL-GEBRA becomes 01 12 07 05 02 18 01. These values are then enciphered as $1^3 \equiv 1 \pmod{101}, 12^3 \equiv 11 \pmod{101}$, etc. The ciphertext becomes

$$01114024087501.$$

These received values are then deciphered by raising each number in the received word to the power $t = 67$ calculated modulo 101.

We now briefly describe the **RSA cryptosystem**, named after its founders R. Rivest, A. Shamir, and L. Adleman; see [30]. This system relies heavily on properties of Euler's ϕ function.

In order to have security, we start with two large primes p and q, which are kept secret by the receiver R. Their product $n = pq$ is publicly announced to all users. For minimal security today, we require that the two primes each contain at least 100 digits. For even more security, we require that the primes p and q be even larger.

The receiver R chooses a positive integer a with a relatively prime to $\phi(n)$. How does R do this? Recall that only R knows the two primes p and q, so she can calculate $\phi(n)$ as

$$\phi(n) = \phi(pq) = \phi(p)\phi(q) = (p-1)(q-1).$$

The receiver R now chooses a positive integer a that is relatively prime to the value $\phi(n)$. In practice the receiver R often randomly chooses a positive integer a and then uses the Euclidean Algorithm (Theorem 1.3) to calculate the greatest common divisor of a and $\phi(n)$. If a is not relatively prime to $\phi(n)$, then the receiver R chooses another value of a and the process is repeated until the receiver finds a positive integer a which is relatively prime to $\phi(n)$.

The integer a is called the **encryption or enciphering exponent**. This encryption exponent a is made public to all users. It will be used to scramble all messages before they are sent.

Also, we must be able to convert our message to a string of numbers. We might use a correspondence something like the following: a will be denoted by 01, b will be denoted by 02, $\ldots$, z will be denoted by 26. We will likely want to have spaces, commas, etc., so we might say that a space will be denoted by 00, a comma by 27, and an ! by 28.

With the above strategy, the message "we love math!" would then become

$$2305001215220500130120 0828.$$

Normally one transmits a fixed number of digits at a time, not the entire string, which may be very long. The only requirement is that the number being transmitted must be relatively prime to the moduls n. This is, however, easily accomplished by simply transmitting a block of digits whose length is less than either of the values p and q. Recall that p and q are very large, so this is easily done.

Let m be a message. Once we have our message m converted to a string of numbers using an identification of letters and numbers as above, how does one encrypt the message before sending it across the wire? We can proceed as follows.

A sender A calculates $m^a \equiv c \pmod{n}$ to obtain the value c, called the **ciphertext**. Thus the value c is simply the least non-negative residue of m^a modulo n, i.e., $c \equiv m^a \pmod{n}$. It represents the scrambled version of the original message.

Upon receiving the value c, how does the receiver R invert this value c to obtain the original message m? This can be done as follows using Euler's Theorem (Theorem 1.30). Since a and $\phi(n)$ are relatively prime, i.e., $\gcd(a, \phi(n)) = 1$, there are integers x and y so that $ax + \phi(n)y = 1$. Note that the value x is simply the inverse of a modulo n.

The receiver R then calculates

$$c^x \equiv (m^a)^x \equiv m^{ax} \equiv m^{1-\phi(n)y} \equiv m(m^{\phi(n)})^{-y} \equiv m(1)^{-y} \equiv m \pmod{n}.$$

This calculation is valid as long as m and n are relatively prime. This will be the case as long as m is smaller than either of the two primes p and q. Recall that the two primes p and q are assumed to be very large, so this will certainly be the case.

Thus Euler's Theorem saves the day by allowing the receiver to invert the received scrambled message c and obtain the original message m!

Example 1.38 *We now illustrate the RSA cryptosystem with a small example. Keep in mind that the numbers are so small that this system does not offer any security. It is for illustration only.*

Assume that $p = 5$ and $q = 11$ so that $n = 5(11) = 55$ is made public, but the receiver R keeps the primes 5 and 11 to himself.

The receiver R then calculates

$$\phi(n) = \phi(55) = \phi(5)\phi(11) = 4(10) = 40.$$

The receiver R now finds a value a, which is relatively prime to $\phi(n)$, i.e., which is relatively prime to 40. He chooses the value $a = 27$ as the enciphering exponent.

Assume that the sender A wants to send the message $m = 8$, so she calculates

$$c \equiv 8^{27} \equiv (2^3)^{27} \equiv 2^{81} \equiv 2^{40}2^{40}2 \pmod{55}.$$

Recall from Euler's Theorem that $2^{40} \equiv 1 \pmod{55}$ and hence that $c \equiv 2 \pmod{55}$.

This ciphertext $c = 2$ is sent to the receiver. Upon receiving the value 2, the receiver needs to have the value x so that $27x + 40y = 1$. Using the Euclidean Algorithm, the receiver is able to calculate that $x \equiv 3 \pmod{40}$ works because $27(3) - 2(40) = 1$. Note that $3(27) \equiv 1 \pmod{40}$. Hence the receiver calculates $2^3 \equiv 8 \pmod{55}$, and hence, he has the original message $m = 8$.

Why does this system give security when sending messages using very large primes? This happens because it is believed, though not yet mathematically proved, that given a value $n = pq$ where p and q are both large primes, it is difficult to compute $\phi(n)$ without knowing the prime factorization of $n = pq$. Thus in the case of the RSA system, everyone knows that n is the product of two primes, but if n is very large, it will be very difficult to determine $\phi(n)$ without knowing the two primes p and q.

Example 1.39 *We now include a larger example to illustrate the RSA system. Even here there is no real security since the numbers are still small, but hopefully this example will serve to further illustrate the details of the RSA system with a somewhat larger example.*

Let $p = 73$ and $q = 151$ so that $n = pq = 11,023$. The reader should first check that both p and q are primes. The receiver calculates

$$\phi(n) = \phi(73(151)) = 72(150) = 10,800.$$

The receiver then chooses an enciphering exponent a that is relatively prime to $\phi(n) = 10,800$; say $a = 11$ is chosen. Now assume that the letters $A, B, C, \ldots, Z$, correspond to $00, 01, 02, \ldots, 25$. Then the message "I LIKE ALGEBRA" becomes the string of digits

$$0811081004001106040117 00.$$

We now break the message into blocks of the same length, say 0811, 0810, 0400, etc.

We encipher these messages as $0811^{11} \equiv 867 \pmod{11023}$, etc. These messages are then sent to our authorized receiver, who knows the values of p and q and can thus determine the deciphering exponent x where x satisfies $ax + \phi(n)y = 1$. After some calculation which you should perform, the receiver finds the deciphering exponent $x = 5891$.

The receiver then calculates $867^{5891} \equiv 811 \pmod{11023}$. He goes through this kind of calculation for each of the $24/4 = 6$ received ciphertexts. Recall that the receiver only needs to calculate the deciphering exponent x once, not once for each received message. Then with x in hand and using the agreed–upon correspondence between letters of the alphabet and numbers, he obtains the original message.

In modern communications, there are also many places where it is important to be able to send messages error-free from one place to another. Contrary to cryptography, in this kind of setting, we are only interested in being able to accurately send messages from one place to another. Developing methods to do this is the study of error-free codes. These codes are often called algebraic codes. The term algebraic is used because we will use various methods from abstract algebra. We will discuss these coding theory ideas in Chapter 8.

1.6 Exercises

1. In an RSA system, assume that $p = 23$ and $q = 29$ so that $n = 667$. Find $\phi(n)$. If $a = 3$ is the encryption exponent, find the decryption exponent x.

2. In the above RSA system, how many different encryption exponents a are there with $1 \le a < n$?

3. In the above RSA system, encipher the message $m = 6$ to obtain the ciphertext $c \equiv m^a \pmod{n}$.

4. Using the encryption system described just before Example 1.36, assume that $n = 17$ and $s = 7$. Encipher the message $m = 2$. What is the deciphering exponent t? Decipher the above–received enciphered message.

5. Using the encryption system described just before Example 1.36, assume that $n = 37$ and $s = 5$. Encipher the message $m = 2$. What is the deciphering exponent t? Decipher the above–received enciphered message.

Chapter 2

Groups

In this chapter we discuss the concept of a group. This is one of the most fundamental concepts in abstract algebra. Many other algebraic structures contain groups.

In subsequent chapters we will see that every ring contains a group and every field (including the real numbers) contains two groups. Vector spaces are examples of algebraic objects that contain a linear structure, and they also always contain a group. Rings, fields, and vector spaces will be studied in detail in Chapters 3, 4, and 6.

It turns out that many algebraic objects with which you are already familiar actually form groups. After looking at numerous examples, we will discuss a variety of properties satisfied by groups.

2.1   Definition of a group

Definition 2.1 *A* **group** *is a non-empty set G with a binary operation $*$ such that the following properties hold:*

1. *Closure: $a * b \in G$ for each $a, b \in G$;*

2. *Associativity: $(a * b) * c = a * (b * c)$ for all $a, b, c \in G$;*

3. *Identity: There is an element $e \in G$, called the* **identity** *element of G, such that $e * a = a * e = a$ for all $a \in G$;*

4. *Inverses: For each $a \in G$ there is an element $a^{-1} \in G$, called the* **inverse** *of a, such that $a * a^{-1} = a^{-1} * a = e$.*

We will often abuse the notation above and refer to the group G (rather than the set G), and suppress the operation $*$ and simply write ab instead of $a * b$.

A group G is **Abelian** or **commutative** if $a * b = b * a$ for all $a, b \in G$. Note that being Abelian is not part of the definition of a group. As we will see, some groups are Abelian and some are not. The term Abelian is used in honor of Niels Abel (1802 – 1829) who was an early pioneer in the study of groups.

Finally, we say that G is a **finite** group if G contains only a finite number of distinct elements; otherwise G is an **infinite** group.

Note that a binary operation is simply a way of combining two elements to obtain a third. For example, on the set of even integers, the operation $*$ defined by $a * b := a + b$ is closed since the sum of two even integers is again even; i.e., $2m + 2n = 2(m + n)$ for any integers m and n. However, on the set of odd integers, the usual operation of addition is not closed, since the sum of two odd integers is even, not odd. Other examples of familiar binary operations include subtraction, multiplication, and division, although as the reader should note, some of these binary operations are not closed. For example, it should be clear that the binary operation of division is not closed over the set of integers (since, for example, $17 \div 5$ is not an integer). However, division is closed over the set of non-zero rational numbers, since dividing one non-zero rational number by a second non-zero rational number still yields a non-zero rational number.

We use $*$ to represent our "generic" binary operation, since in some cases the group operation will be addition, in some cases it will be multiplication, and in still other cases it may be a rather unusual operation. We will see such examples soon.

When using the notation a^{-1}, we need to be careful, as this does not mean to divide 1 by the element a; a^{-1} is just the notation used to represent the inverse of the element a with respect to the group operation $*$. Thus, if $*$ represents addition of integers, then a^{-1} will actually turn out to be $-a$ since, in this case, the identity e is 0, and

$$a + (-a) = (-a) + a = 0$$

for all integers a.

As will be indicated in Exercise 2.2.3, the identity e in a group is unique, and for each element a in a group, we have a unique inverse a^{-1} in the group; see Exercise 2.2.4.

2.2   Examples of groups

1. The simplest group is the set $G = \{e\}$, which consists of just one element e. We define the binary operation $*$ by $e * e := e$. The reader should check that G forms a finite Abelian group, albeit a rather trivial one.

2. One of the best–known examples of a group is the set $\mathbb{Z}$ of integers under the usual operation of addition. This operation is clearly closed since the sum of two integers is an integer. Moreover, $e = 0$ is the identity, and for an integer a, $a^{-1} = -a$ since the operation is addition. This is a natural example of an infinite group, that is, a group with an infinite number of elements.

3. As another example of an additive group, we can consider the set $\mathbb{Z}_n$ of integers $\{0, 1, \ldots, n-1\}$ and define addition modulo n. The reader should check that $\mathbb{Z}_n$, endowed with the operation of addition modulo n, forms a commutative group.

4. Similarly for multiplication, one can consider the set $\mathbb{Z}_n^*$ consisting of all invertible elements from $\mathbb{Z}_n$. Recall from Chapter 1 that an element $a \in \mathbb{Z}_n$ is invertible if and only if $\gcd(a, n) = 1$, i.e., if and only if a and n are relatively prime. If we define an operation $a * b$ to be the product ab calculated modulo n, then we have a commutative group, as the reader should verify.

   By way of illustration, in the group $\mathbb{Z}_5^*$ of non-zero integers modulo 5 under multiplication, the element 3 is invertible since $3 \cdot 2 \equiv 1 \pmod{5}$; see Theorem 1.35. Note that the inverse of 3 is the element 2 and the inverse of 2 is the element 3. In fact, each of the non-zero values $1, 2, 3, 4$ is invertible modulo 5 since each of these values is relatively prime to 5.

   In the set $\{1, 2, 3, 4, 5\}$ of non-zero integers modulo 6, only the elements $1, 5$ are invertible since $\gcd(1, 6) = \gcd(5, 6) = 1$ (and the inverse of 1 is 1 while the inverse of 5 is 5 modulo 6). In contrast, $\gcd(2, 6) = 2$, $\gcd(3, 6) = 3$, and $\gcd(4, 6) = 2$. Hence, $2, 3, 4$ are not invertible modulo 6.

5. We can often illustrate the group operation via a group table where, at the intersection of row x and column y, we place the group element $x * y$. We illustrate with the group $\mathbb{Z}_5$ of integers under the operation of addition modulo 5.

+	0	1	2	3	4
0	0	1	2	3	4
1	1	2	3	4	0
2	2	3	4	0	1
3	3	4	0	1	2
4	4	0	1	2	3

6. We now give the group table for the group $\mathbb{Z}_8^*$ of invertible elements modulo 8. Here the group $\mathbb{Z}_8^* = \{1, 3, 5, 7\}$ since for each of these elements a, we have $\gcd(a, 8) = 1$.

×	1	3	5	7
1	1	3	5	7
3	3	1	7	5
5	5	7	1	3
7	7	5	3	1

7. One can also use polynomials to form various kinds of groups. As an illustration, consider the set G of all polynomials with real coefficients. We can make G into a group by adding polynomials using ordinary polynomial arithmetic. For example, the identity polynomial e for this group will be the constant polynomial $e = f(x) = 0$ and the inverse of a polynomial $f(x) = \sum_{i=0}^{n} f_i x^i$ will be the polynomial $-f(x)$ defined by

$$-f(x) = \sum_{i=0}^{n} -f_i x^i.$$

Again, the reader should check that with this operation, the set G forms a group which is, in fact, also commutative (and infinite).

8. Similarly one can form a group G by considering all polynomials with real coefficients whose degrees are at most n. The operation is again just the usual addition of polynomials.

9. Consider the Cartesian product $G = \mathbb{Z}_2 \times \mathbb{Z}_2$ of the integers modulo two with itself. We can make G into a group by defining

$$(a, b) + (c, d) = (a + c, b + d)$$

for any $a, b, c, d \in \mathbb{Z}_2$. The reader should check that with this binary operation, G is an Abelian group containing four distinct elements:

$$(0,0), (0,1), (1,0), (1,1).$$

For more details regarding the Cartesian product, see Section 3 of the Appendix.

10. We now consider rotations of an equilateral triangle that will be seen to form a group with six elements. Consider the triangles that appear below.

We can rotate the triangle counter-clockwise about its mid-point by an angle of $2\pi/3$ radians or 60 degrees. In the following discussion we will denote this operation by ρ. We will denote by R the operation of reflecting the triangle about a vertical line from the top of the triangle to the mid-point of the base. For the equilateral triangle, there are six different operations, which can be denoted by $e, \rho, \rho^2, R, \rho R$, and $\rho^2 R$. The identity is denoted by e. There are, of course, other operations, but they can be derived from these six operations.

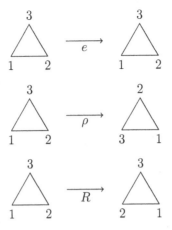

These six operations form a non-commutative group as described in the group table.

	e	ρ	ρ^2	R	ρR	$\rho^2 R$
e	e	ρ	ρ^2	R	ρR	$\rho^2 R$
ρ	ρ	ρ^2	e	ρR	$\rho^2 R$	R
ρ^2	ρ^2	e	ρ	$\rho^2 R$	R	ρR
R	R	$\rho^2 R$	ρR	e	ρ^2	ρ
ρR	ρR	R	$\rho^2 R$	ρ	e	ρ^2
$\rho^2 R$	$\rho^2 R$	ρR	R	ρ^2	ρ	e

We note that this group is not Abelian, since the group table is not symmetric about the main diagonal. In particular, note that $R\rho \neq \rho R$ since $R\rho = \rho^2 R$, which is not the same as ρR.

11. Suppose that we now consider a similar situation except that we replace the triangle by a square and once again consider symmetries. We can let ρ denote the rotation about the center of the square by $2\pi/4$ radians and let R denote the reflection in the perpendicular bisector of the side that joins vertices 1 and 2. If you do some calculating with the various symmetries, you will find that we now obtain a group with eight distinct elements. One can check that $\rho^4 = e$, $R^2 = e$, and $\rho^3 R = R\rho$. This group is often called the **dihedral group** of order 8.

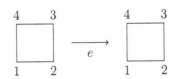

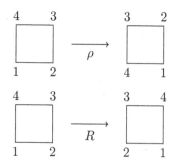

We now include the table for this group.

	e	ρ	ρ^2	ρ^3	R	ρR	$\rho^2 R$	$\rho^3 R$
e	e	ρ	ρ^2	ρ^3	R	ρR	$\rho^2 R$	$\rho^3 R$
ρ	ρ	ρ^2	ρ^3	e	ρR	$\rho^2 R$	$\rho^3 R$	R
ρ^2	ρ^2	ρ^3	e	ρ	$\rho^2 R$	$\rho^3 R$	R	ρR
ρ^3	ρ^3	e	ρ	ρ^2	$\rho^3 R$	R	ρR	$\rho^2 R$
R	R	$\rho^3 R$	$\rho^2 R$	ρR	e	ρ^3	ρ^2	ρ
ρR	ρR	R	$\rho^3 R$	$\rho^2 R$	ρ	e	ρ^3	ρ^2
$\rho^2 R$	$\rho^2 R$	ρR	R	$\rho^3 R$	ρ^2	ρ	e	ρ^3
$\rho^3 R$	$\rho^3 R$	$\rho^2 R$	ρR	R	ρ^3	ρ^2	ρ	e

12. The previous two examples suggest a whole class of groups described by symmetries of a regular n-sided polygon. For an n-sided regular polygon, the group has $2n$ distinct elements and can be generated by the rotation ρ counter-clockwise about the center of the polygon, by $2\pi/n$ radians, together with a reflection R in the perpendicular bisector of any of the sides. These operations are subject to $\rho^n = e$, $R^2 = e$, $\rho^{n-1}R = R\rho$.

13. Consider a "real-world example" of a group of symmetries. Assume that we have a rectangular bed mattress that has been in position on its frame for some time. How can the mattress be lifted and placed back on the stationary frame?

  • We can simply place the mattress back as it was originally;

  • We can place the foot of the mattress at the head of the bed and the head of the mattress at the foot of the bed (think of this as a rotation);

  • We can flip the mattress over leaving the head of the mattress at the head of the bed and the foot of the mattress at the foot of the bed;

  • We can rotate and flip the mattress.

These four actions form a group, sometimes known as the "mattress group" for obvious reasons.

The next several examples of groups deal with matrices. If you are not familiar with basic matrix operations, please consult the Appendix where in Section 6 we discuss some basic properties of matrices.

14. Consider the set G of all $n \times n$ matrices whose entries are real numbers. Let $A = (a_{ij})$ and $B = (b_{ij})$ be two $n \times n$ matrices in G. Then we can define a binary operation $*$ on G by adding the elements of A and B that lie in the same position. More formally, we can define

$$A * B := A + B = (a_{ij}) + (b_{ij}) = (a_{ij} + b_{ij}), 1 \leq i, j \leq n.$$

We note that this operation is closed, the identity e will be the $n \times n$ matrix consisting of all zeros, and the additive inverse of A will be the matrix $-A = -(a_{ij}) = (-a_{ij})$. You should convince yourself that with this operation, G forms a commutative group.

15. What about $n \times n$ matrices under the usual multiplication of matrices? Do they form a group? No, since not every matrix has a multiplicative inverse. For example, the matrix all of whose entries are 0 does not have a multiplicative inverse. If, however, we restrict our attention to those matrices that are **non-singular**, i.e., those matrices whose determinant is not zero (and which thus have rank n), then we obtain a group. However, this is not a commutative group since, in general, $AB \neq BA$. For example, if

$$A = \begin{bmatrix} 1 & 2 \\ 3 & 4 \end{bmatrix} \text{ and } B = \begin{bmatrix} 5 & 6 \\ 7 & 8 \end{bmatrix},$$

then

$$AB = \begin{bmatrix} 19 & 22 \\ 43 & 50 \end{bmatrix} \text{ while } BA = \begin{bmatrix} 23 & 34 \\ 31 & 46 \end{bmatrix}.$$

In this group the identity e will be the $n \times n$ identity matrix, i.e., the matrix with n ones on the main diagonal and zeros elsewhere. The $n \times n$ identity matrix is often denoted by I_n in the mathematical literature.

Hopefully most of the above examples will seem to the reader to be centered in rather natural settings involving integers, polynomials, and matrices. We now give another example of a group that may seem to be rather unnatural.

16. Let $G = \mathbb{Z}$ denote the set of integers. For two integers a and b, we define a binary operation by $a * b := a + b + 2$. Does this provide another example of a group? Let's check. First the operation is closed since if a, b are integers, so is $a + b + 2$. For associativity, given integers a, b, c we calculate as follows. First,

$$(a * b) * c = (a + b + 2) * c = (a + b + 2) + c + 2 = a + b + c + 4.$$

Similarly, we have

$$a * (b * c) = a * (b + c + 2) = a + (b + c + 2) + 2 = a + b + c + 4.$$

So the operation $*$ is associative.

The identity e will have to satisfy $a * e = a$ so that

$$a + e + 2 = a.$$

This means that $e = -2$. Well, -2 is an integer, so it is in $\mathbb{Z}$ and we have

$$a * (-2) = a + (-2) + 2 = a,$$

as desired. Similarly $(-2) * a = a$.

Finally for inverses, we are required to have $a * b = -2$ now that we know $e = -2$, i.e., $a + b + 2 = 2$ in order for the integer a to have an inverse b. Thus, if we take $b = -a - 4$, we have

$$a * (-a - 4) = a + (-a - 4) + 2 = -2,$$

as needed. Also $(-a - 4) * a = -2$ as required. Therefore, $a^{-1} = -a - 4$ in this case.

Thus, the integers with this rather unusual operation do indeed form a group. Is the group commutative? Sure, since

$$a * b = a + b + 2 = b + a + 2 = b * a.$$

Is there anything special about 2? What if we define $a * b = a + b + k$ where k is a fixed integer? Do we still obtain a group for each integer k? The answer is yes! Note that if $k = 0$, we have the group $\mathbb{Z}$ of integers with the usual addition operation.

17. Earlier we saw an example of a matrix group that was non-Abelian: in particular, the set of $n \times n$ non-singular matrices under the usual operation of matrix multiplication. We now give another example of a non-Abelian or non-commutative group.

We will construct this group using permutations, so we ask the reader to review the corresponding material related to permutations in the Appendix, Section 5.

Let G be the set of the $3! = 6$ permutations on the set $X = \{1, 2, 3\}$. In cycle notation as discussed in Section 5 of the Appendix,

$$G = \{(1)(2)(3), (12)(3), (13)(2), (1)(23), (123), (132)\}.$$

For $f, g \in G$, we define our operation $\circ$ on G by $(f \circ g)(x) := f(g(x))$, i.e., we define our operation by composition of permutations. The reader should check that with this operation, G is a group.

Recall that when computing with permutations, we always work from right to left. As an illustration, note that

$$(12)(3) \circ (123) = (1)(23).$$

We also note that

$$(123) \circ (12)(3) = (13)(2),$$

which is not the same permutation. Hence this group is not commutative.

This group G is often called the **symmetric group** on three elements and is often denoted by S_3. In general, one can also form a symmetric group S_n on n symbols where n is any positive integer by using all of the permutations on n symbols, of which there are $n!$ distinct permutations. For each $n \geq 3$, S_n is not a commutative group.

18. In the symmetric group S_3, in cycle notation, consider the set

$$A_3 = \{(1)(2)(3), (123), (132)\}$$

of three permutations. One can show that A_3, with the operation of composition of permutations, forms a group of its own (sitting inside S_3). This group is usually called the **alternating group** A_3 on three elements. For those readers familiar with the theory of permutations, these are even permutations, i.e., each permutation has a sign that is equal to 1.

In general, for any positive integer n, the set of even permutations forms a subgroup of S_n, called the **alternating group** A_n.

Such a group, which "sits inside" a larger group, is called a **subgroup** of the larger group. Subgroups are the subject of the next section of this chapter.

2.2 Exercises

1. Build the addition group table for the integers $\mathbb{Z}_m$ modulo m for each $2 \leq m \leq 10$.

2. Build the multiplication group table for the invertible integers $\mathbb{Z}_m^*$ modulo m for each $2 \leq m \leq 10$.

3. Show that in a group, the identity element e must be unique; i.e., a group has only one identity element.

4. Show that the inverse a^{-1} of an element a in a group must be unique.

5. Assume that G is a group in which $a^2 = e$ for all elements in the group G. Show that the group G must be Abelian.

6. If G is a group and $a, b \in G$, show that

$$(ab)^{-1} = b^{-1}a^{-1}.$$

The result of this exercise explains why

$$(AB)^{-1} = B^{-1}A^{-1}$$

where A and B are invertible matrices; see Section 6 of the Appendix.

7. Give an example of a group G with elements a and b in G so that

$$(ab)^{-1} \neq a^{-1}b^{-1}.$$

8. Show that if a, b are elements in a group G, then the equation $ax = b$ has a unique solution $x \in G$. Similarly show that $xa = b$ also has a unique solution $x \in G$.

9. If $a, b \in G$, a group with operation $*$, and $a * b = a * c$, prove that $b = c$. Does the same conclusion hold if $a * b = c * a$?

10. Explain why the integers under the operation of subtraction do not form a group.

11. Let G be a group. For elements $a, b \in G$, define a to be related to b if there is an element $g \in G$ so that $b = gag^{-1}$. Show that this relation is an equivalence relation on the group G, i.e., show that this relation is reflexive, symmetric, and transitive. See the Appendix, Section 3, for more information about equivalence relations.

12. Let G be a group and assume that b, c, d, e, f are elements of G. Solve for y in the group equation $byd = e$ and in the equation $cydf = bc$.

13. In the group $\mathbb{Z}$ of integers under addition, define

$$a * b = a + b + 1.$$

Show that with the operation $*$, $\mathbb{Z}$ forms a group. Solve the equation $50 * x = 100$ for x.

14. What is the product of all the elements of a finite commutative group? Can the same conclusion be drawn if the group is not commutative?

15. Let G be a group and let g, h be elements of G. Show that, for all positive integers n,

$$(g^{-1}hg)^n = g^{-1}h^n g.$$

16. Show that $\{1, -1, i, -i\}$ with ordinary complex number multiplication is a group. See Section 7 of the Appendix for details of complex number multiplication.

17. Let $(G, *)$ and $(H, \circ)$ be two groups. Consider the Cartesian product set $G \times H$. Show that, if we define a binary operation $\triangle$ on $G \times H$ by

$$(g_1, h_1)\triangle(g_2, h_2) = (g_1 * g_2, h_1 \circ h_2),$$

the set $G \times H$ forms a group.

Note that if one takes $G = \mathbb{R}$ to be the group of real numbers, then the Cartesian product $G \times G = \mathbb{R} \times \mathbb{R} = \mathbb{R}^2$ is simply the xy-plane so often discussed in calculus.

18. An automorphism f of a group G is a mapping from G to itself which is 1-1, onto, and has the property that for all $a, b \in G$

$$f(a * b) = f(a) * f(b).$$

Let G be a group with an operation $*$. Show that the map $f : G \to G$ defined by $f(a) = a^{-1}$ for each $a \in G$ is a group automorphism if and only if the group is Abelian.

2.3 Subgroups

Some groups possess a very rich "internal structure." Indeed, as we noted briefly in the previous section, some groups have smaller groups inside of them. We now study such situations in detail and begin with the following definition.

Definition 2.2 *A non-empty subset H of a group G with operation $*$ is a* **subgroup** *of G if H is itself a group under the same operation $*$ as in G.*

Thus, a subgroup must contain an identity element; let's call this identity element f. Then
$$f * f = f = e * f$$
where e is the identity of the original group G. But since f must have an inverse in the group G, we see that $f = e$. Thus the identity in the subgroup H must be the same as the identity in the group G.

It might appear from the definition that, in order to check whether a subset H of a group G is a subgroup, we have to check the four properties required for a group as given in Section 2.1. However, the following result shows that we can simply check either two conditions as in part (ii), or even one slightly more complicated condition, as in part (iii), in order to confirm that a subset H of G is actually a subgroup of G.

Theorem 2.3 *The following conditions on a non-empty subset H of a group G are equivalent; i.e., each of the three conditions (i), (ii), and (iii) implies each of the others:*

    (i) H is a subgroup of the group G;

    (ii) H satisfies

      (a) if $h \in H$ then $h^{-1} \in H$ and

      (b) if $h, k \in H$ then $h * k \in H$;

    (iii) H satisfies if $h, k \in H$ then $h * k^{-1} \in H$.

Proof: We will show that part (i) implies part (ii), part (ii) implies part (iii), and that part (iii) implies part (i) to complete the proof.

Part (i) clearly implies part (ii) since the properties in part (ii) are properties that must hold in a group, and hence in a subgroup.

To prove part (iii) assuming part (ii), let $h, k \in H$. Then part (ii)(a) implies that $k^{-1} \in H$ and hence by part (ii)(b), we have that the product $h * k^{-1} \in H$. Thus, part (iii) holds.

We now assume that part (iii) holds and show that part (i) holds.

The operation $*$ is associative in H since it is associative in the larger group G.

If $h \in H$, then using part (iii) with $k = h$, we see that

$$h * k^{-1} = h * h^{-1} = e \in H,$$

so the subgroup H has an identity e.

Now let $h = e$ and let g be an element of H. Then by part (iii) with $k = g$ we have $h * k^{-1} = e * g^{-1} = g^{-1} \in H$.

Finally, let $a, b \in H$. We need to show that $ab \in H$ in order for H to be closed. Using part (iii) with $h = a$ and $k = b^{-1}$ we have that $h * k^{-1} = ab \in H$. Here we have used the fact that $(a^{-1})^{-1} = a$. ∎

There are many examples of subgroups. Perhaps one of the easiest to understand is to let $G = \mathbb{Z}$ be the group of integers under the usual operation of addition, and choose H to be the subset of even integers. In order for H to be a subgroup of the group G, the operation on H must be the same as the operation on G, i.e., it must be addition. Hence by condition (ii)(a), we see that if $h = 2m$ is an even integer, then so is

$$h^{-1} = -h = -2m = 2(-m).$$

Since the operation is addition, the inverse of h must be the negative of the even integer h.

Similarly, by part (ii)(b), if $h = 2m$ and $k = 2n$, then

$$h * k = 2m + 2n = 2(m + n)$$

is even and so is in H.

One could alternatively use condition (iii) to show that H is a subgroup by checking that if $h = 2m$ and $k = 2n$, then

$$h * k^{-1} = 2m + (-2n) = 2m - 2n = 2(m - n)$$

is even and we are done.

What if one considers the set S of odd integers? Will this set S form a subgroup of $G = \mathbb{Z}$? The answer is no for at least two reasons. First, the identity of S must be the same as the identity for G, which is $e = 0$. But 0 is not odd since $0 \neq 2m + 1$ for any integer m.

One could also show that S is not a subgroup by showing the binary operation of addition is not closed on S. This follows from the fact that the sum of two odd integers is even. For example

$$(2m + 1) + (2n + 1) = 2(m + n + 1)$$

is even, not odd. Thus the subset S is not a subgroup of the group $\mathbb{Z}$ of integers under the operation of integer addition.

If the group G is finite, i.e., if G contains only a finite number of elements, then condition (iii) from Theorem 2.1 can be replaced by the following much simpler condition.

Theorem 2.4 *If G is a finite group with operation $*$, then a non-empty subset H is a subgroup of G if and only if H is closed; i.e., if and only if whenever $h, k \in H$, then $h * k \in H$.*

Proof: We know that $*$ is associative on the subset H since $*$ is associative on the larger group G.

Assume that $H = \{h_1, \dots, h_n\}$. Let $h \in H$ and consider the elements $hh_1, \dots, hh_n$, which are distinct in G; for otherwise, if $hh_i = hh_j$, then $h_i = h_j$.

Since $h \in H$, we must have in the group G that $h = hh_j$ for some $j = 1, \dots, n$. Hence multiplying by h^{-1} we have $e = h_j$ so that the subset H contains an identity e.

Since $e = hh_k$ for some k, h_k looks like a potential inverse for the element h. In the group G we must have that $hm = mh = e$ for some group element m. Thus in G we have $e = hh_k = hm$. But by Exercise 2.2.8, in any group the equation $a = bx$ has a unique solution. Therefore $h_k = m = h^{-1}$ and the proof is complete. ∎

We now give a way to generate several subgroups within a given group.

Definition 2.5 *Let G be a group with operation $*$ and let g be an element in G. Then the **subgroup generated by** g is the set $\langle g \rangle$ of all powers of g, i.e.,*

$$\langle g \rangle = \{g^n \mid n \text{ is an integer}\}.$$

*The element g is a **generator** for the subgroup $\langle g \rangle$.*

Here by g^n we mean the group element

$$\underbrace{g * g * \cdots * g}_{n \text{ copies of } g}$$

where we operate on g with itself n times. Note that we are calling this a "subgroup" generated by g so it better be a subgroup of the group G! This is indeed the case, and we ask the reader to provide a proof in the exercises.

We now briefly discuss some properties of powers of group elements. As the reader will quickly detect, in many respects, these powers of group elements behave similarly to powers of real numbers.

Let g, h be elements in a group G. Then for any integers r and s we have
(i) $g^r g^s = g^{r+s}$;
(ii) $(g^r)^s = g^{rs}$;
(iii) $g^{-r} = (g^r)^{-1} = (g^{-1})^r$;
(iv) if $gh = hg$, i.e., if the elements g and h commute, then $(gh)^r = g^r h^r$.

We note that property (iv) only holds when the elements g and h commute with each other.

We now give several examples to illustrate these ideas. Consider the multiplicative group $\mathbb{Z}_5^* = \{1, 2, 3, 4\}$ of invertible integers modulo 5. We see that

$$2^6 \equiv (2^2)^3 \equiv 4^3 \equiv 64 \equiv 4 \pmod 5.$$

Similarly,

$$2^{100} \equiv (2^4)^{25} \equiv 16^{25} \equiv 1^{25} \equiv 1 \pmod 5.$$

For negative exponents we may proceed in a similar fashion. For example,

$$3^{-12} \equiv (3^{12})^{-1} \equiv (3^4)^{-3} \equiv 81^{-3} \equiv 1^{-3} \equiv 1 \pmod 5.$$

On the other hand, we could have also found this by calculating

$$3^{-12} \equiv (3^{-1})^{12} \equiv 2^{12} \equiv (2^4)^3 \equiv 1^3 \equiv 1 \pmod 5.$$

We note that property (iii) indicates that g raised to a negative power can be interpreted in one of two ways; namely as the inverse of g raised to the corresponding positive power, or as the corresponding positive power of the element g raised to the power -1. Thus in any case we have

$$g^{-r} = (g^{-1})^r = (g^r)^{-1}.$$

The comments above regarding powers of group elements lead to a very natural type of group.

Definition 2.6 *A group G is* **cyclic** *if there is an element $g \in G$ so that $G = \langle g \rangle$. In such a case g is called a* **generator** *for the cyclic group G.*

Perhaps the best example of a cyclic group is the group $\mathbb{Z}$ of integers under addition. Here $g = 1$ is a generator since if n is a positive integer,

$$g^n = \underbrace{1 + 1 + \cdots + 1}_{n \text{ copies of } 1} = n,$$

and $g^0 = 1$, while if n is negative, then $g^n = -n$.

In the group $G = \mathbb{Z}_5^*$ of non-zero integers with the operation of multiplication modulo 5, then $g = 2$ and $g = 3$ are generators. However, $g = 1$ and $g = 4$ are not generators for the group G. For example, $2^0 = 1, 2^1 = 2, 2^2 = 4$ and $2^3 = 3$ where all calculations are done modulo the prime $p = 5$. This shows that 2 is a generator of this group because each of the elements $\{1, 2, 3, 4\}$ is "generated" by a power of 2.

In the exercises we ask the reader to prove that every cyclic group is Abelian.

Next, we look at the structure of the subgroups of a cyclic group.

Theorem 2.7 *Every subgroup of a cyclic group is cyclic.*

Proof: Let G be a cyclic group and let H be a subgroup of G. If $H = \{e\}$ consists of just one element, then H is clearly cyclic. Thus, we may assume there is an element $h \in H$ such that $h \neq e$.

Since the group G is cyclic, G has a generator, say $g \in G$. Hence we may assume that $h = g^n$ for some positive integer n. By the Well Ordering Principle (see the Appendix, Section 2), let m be the **smallest** positive integer so that $g^m \in H$. We claim that $H = \langle g^m \rangle$.

Let k be an arbitrary element in H that is, of course, also in G. Thus, k can be written as $k = g^s$ for some positive integer s.

We now apply the Division Algorithm to write $s = qm + r$ where $q \geq 0$ and $0 \leq r < m$. Hence we have that

$$k = g^s = g^{qm+r} = (g^m)^q g^r$$

using the properties of powers outlined above. Note that k and g^m are in H, which is a subgroup. Thus since $m < r$, it must be the case that r is forced to be 0. Hence

$$k = (g^m)^q g^0 = (g^m)^q.$$

Thus, k is a power of g^m, meaning that g^m is a generator of H. Therefore, H must be cyclic; that is, $H = \langle g^m \rangle$. ∎

Remark 2.8 *Even if it is known that a group is cyclic, it may be difficult to find a generator. For example, the number 4999 is a prime (you should check this!) and it is known that there are exactly $\phi(4998) = 1344$ generators of the group $\mathbb{Z}_{4999}^*$ of invertible integers modulo 4999 under the operation of multiplication modulo 4999. It is, however, not easy to find one generator even though almost 27% of the elements in the group $\mathbb{Z}_{4999}^*$ are generators. Recall from Chapter 1 that ϕ denotes the Euler function.*

2.3 Exercises

1. Show that the intersection of two subgroups of a group is itself a subgroup of the group. Show that a similar result holds for the intersection of any finite number of subgroups of a group.

2. Determine whether the union of two subgroups of a group is always a subgroup of the group.

3. Find four distinct subgroups of the group $\mathbb{Z}$ of integers under addition.

4. If G is an Abelian group, show that the set $\{x|x = x^{-1}\}$ is a subgroup of G.

5. Let $k \geq 2$ be an integer. Show that $\{kn|n \in \mathbb{Z}\}$ is a subgroup of $\mathbb{Z}$ under the usual addition operation.

6. Let Q^* denote the set of all non-zero rational numbers. This set along with the usual operation of multiplication forms a commutative group. Let Q^+ denote the subset of Q^* of positive rational numbers. Is Q^+ a subgroup of Q^*?

7. Let G be a group and fix $a \in G$. Consider the set

$$N_a = \{x|xa = ax\}.$$

Note that the set N_a is the set of all elements in the group that commute with the element a. Show that N_a is a subgroup of G.

8. Show that in a group G, the set

$$\langle g \rangle = \{g^n|n \in \mathbb{Z}\}$$

of powers of the element $g \in G$ is indeed a subgroup of G.

9. Show that every cyclic group is Abelian.

10. In an infinite cyclic group, show that every subgroup except $\{e\}$ must be infinite.

11. The **center** $Z(G)$ of a group G is the set of all elements in G that commute with every element of G. In symbols,

$$Z(G) = \{g \in G|gh = hg \text{ for all } h \in G\}.$$

For any group G, explain why $Z(G)$ is not empty. What is $Z(G)$ if G is Abelian? Finally, show that the center of any group G is always a subgroup of the group G.

12. Define the product AB of two subsets A and B of a group G by

$$AB = \{ab|a \in A, b \in B\}.$$

Show that $HH = H$ if and only if H is a subgroup of G.

2.4   Cosets and Lagrange's Theorem

In this section we define cosets and study some of their properties. We then prove a famous theorem of Lagrange which is, arguably, the most important theorem in the study of finite groups.

Definition 2.9 *Assume that H is a subgroup of a group G with operation $*$. Let a be an element of the group G (which is not necessarily an element of H). Define the **left coset** $a * H$ of a with respect to H by*

$$a * H = \{a * h | h \in H\}.$$

*One can also define the **right coset** of a with respect to H by $H * a = \{h * a | h \in H\}$.*

There is no real reason to consider right cosets over left cosets (if the group G is Abelian, they are the same), so from here on we only discuss left cosets. In fact, we will usually just talk about cosets with the understanding that they are left cosets.

Perhaps without realizing it, we have already encountered cosets. Let $G = \mathbb{Z}$ be the group of integers under the usual operation of addition of integers. Let H be the set of even integers. From our earlier discussion, we know that H is a subgroup of G. Let a be an odd integer and consider the coset $a + H = \{a + h | h \in H\}$. We note that, for any even $h \in H$ with $h = 2m$ for some integer m, and $a = 2n + 1$, we have

$$a + h = 2n + 1 + 2m = 2(m + n) + 1.$$

This is clearly odd. Thus the coset $a + H$ is simply the set of odd integers. We also note that if a is an odd integer, $H \cap aH = \emptyset$ so the set $\mathbb{Z}$ of integers is disjointly partitioned into two cosets, namely the even integers and the odd integers.

Example 2.10 *Let $G = \{0, 1, \ldots, 7\}$ be the group of integers with the operation being addition modulo 8. The reader should check that the subset $H = \{0, 2, 4, 6\}$ forms a subgroup of G. There are two cosets of G, namely the subgroup H and the coset*

$$1 + H = \{1, 3, 5, 7\}.$$

Notice that the union of these two cosets gives the entire group G.

The reader should also note that there are other cosets, for example $3 + H, 5 + H$, and $7 + H$, but these all contain the same elements as the coset $1 + H$. Similarly $H = 2 + H = 4 + H = 6 + H$.

We note that the subset $K = \{0, 4\}$ is also a subgroup of the group G. Here the disjoint cosets are $K, 1 + K, 2 + K, 3 + K$ and again, the union of these four cosets is the entire group G.

We now highlight some properties of cosets of a subgroup H of a group G.

Theorem 2.11 *Let G be a group and let H be a subgroup of G.*

1. *A subgroup H of G is itself a coset;*

2. *Any element $a \in G$ is in the coset $a * H$;*

3. *If $b \in a * H$, then the coset $b * H = a * H$;*

4. *Unless $a \in H$, the coset $a * H$ is not a subgroup of G;*

5. *If $a, b \in G$, then either $a * H = b * H$ or $(a * H) \cap (b * H) = \emptyset$;*

6. *Any two cosets of H have the same cardinality; and if the group G is finite, they have the same number of elements, which is the number of elements in the subgroup H.*

Proof: We now prove parts (3), (5), and (6) of this theorem, leaving proofs of the remaining parts to the exercises. Throughout this proof, we will, for the sake of simplicity, denote $h * k$ by hk and $a * H$ by aH.

For part (3), assume that $b \in aH$. Then $b = ah$ for some $h \in H$. Let $x \in bH$. Then for some $k \in H$ we have

$$x = bk = (ah)k = a(hk) \in aH,$$

since $hk \in H$ (recall that H is a subgroup so it is closed). Thus we have proved the set inclusion $bH \subseteq aH$.

Now let $y \in aH$. Recall that $b = ah$; this means $a = bh^{-1}$. Then, for some $m \in H$, we have

$$y = am = bh^{-1}(m) = b(h^{-1}m) \in bH$$

so that $aH \subseteq bH$. Hence we have shown that $aH = bH$.

For part (5), if $aH \cap bH \neq \emptyset$, then there is an element $c \in aH$ and $c \in bH$. By part (3), $cH = aH = bH$ so that $aH = bH$, as desired.

For part (6), let $f : H \to aH$ be defined by $f(h) = ah$. The reader should check that the function f is both 1-1 and onto. Thus, it is a bijection. Hence the sets H and aH can be put in a 1-1 correspondence and thus they have the same cardinality. (See the discussion of cardinality of sets in the Appendix, Section 4.) Moreover, if the group G is finite, then the cosets have the same number of elements. ∎

The reader should check these properties in the above examples.

Definition 2.12 *If G is a finite group, then the **order** of G is defined to be the number of distinct elements in G. We denote this by $|G|$.*

We are now ready to state a very important result in the theory of finite groups. This is Lagrange's Theorem, named after its discoverer Joseph L. Lagrange (1736 – 1813).

Theorem 2.13 *(Lagrange's Theorem) Let G be a finite group and let H be a subgroup of G. Then the order $|H|$ of H divides the order $|G|$ of G. More specifically, if the subgroup H has m distinct cosets in G, then*

$$|G| = m|H|.$$

Remark 2.14 *Note that the number of cosets in G must be finite since the group G itself contains only a finite number of elements.*

Proof: We first note that the m cosets of H must partition G into disjoint subsets because from part (5) of Theorem 2.11, any two cosets are either disjoint or identical. Hence we have the **disjoint** set partition of the group G into cosets

$$G = a_1 H \cup \cdots \cup a_m H.$$

for some elements $a_1, a_2, \ldots, a_m \in G$.

From part (6) of Theorem 2.11 we know that, for each coset $a_i H$, we have

$$|a_i H| = |H|.$$

Hence, from the equation above, we have that

$$|G| = m|H|,$$

which completes the proof. ∎

Thus, in a nutshell, Lagrange's Theorem states that the number of elements in a subgroup of a finite group must divide the number of elements in the group itself. We now give several applications of Lagrange's Theorem.

Corollary 2.15 *If G is a group with a prime number of elements, then G must be cyclic.*

Proof: Let $g \neq e$ be an element in G. Clearly G must have such an element since G has a prime number $p \geq 2$ of elements. Then the element g generates a subgroup $H = \langle g \rangle$ of G. By Lagrange's Theorem, the number of elements in H must divide p, the number of elements in G. Since p is a prime, the number of elements in H must be p (the number of elements in H is not 1, since $g \neq e$), so the number of elements in H must be the same as the number of elements in G. Thus $G = H$, so G is cyclic since $H = \langle g \rangle$ is cyclic. ∎

Corollary 2.16 *If g is an element in a finite group G, then the order of the element g (the least positive integer k so that $g^k = e$) must divide the number of elements in the group G.*

Corollary 2.17 *(Fermat's Theorem) If p is a prime and p does not divide an integer a, then $a^{p-1} \equiv 1 \pmod{p}$.*

Corollary 2.18 *(Euler's Theorem) If a and n are relatively prime, then $a^{\phi(n)} \equiv 1 \pmod{n}$.*

Note that the comments above also imply that, in a group of prime order, any non-identity element is a generator of the full group. The reader should contrast this with the remarks at the end of Section 2.3 that discuss the difficulty of locating a generator in a large cyclic group.

Lagrange's Theorem can be used to prove that subgroups of various orders cannot exist in a given group. For example, if G is a group of order 84, G cannot have a subgroup of order 37 since 37 does not divide 84. Similarly, such a group G cannot have subgroups of orders 16, 20, or 60.

We now discuss the order of an element in a group. If $g \in G$, the **order** of g is the least positive integer k with the property that $g^k = e$. Recall g^k means to use the binary group operation $*$ on g, k times; i.e., we calculate

$$\underbrace{g * g * \cdots * g}_{k \text{ copies of } g}.$$

For example, in the group $\mathbb{Z}_5^*$ of integers modulo 5 under multiplication, we note that the element 4 has order 2 since $4^2 \equiv 1 \pmod 5$. We also have that $4^4 \equiv 1 \pmod 5$ but the order of 4 is 2, not 4, since 2 is the **least** positive integer n so that $4^n \equiv 1 \pmod 5$.

Remark 2.19 *One must be careful not to read too much into Lagrange's Theorem. For example, if the order of a finite group G is, say, n, and m divides n, then the group G does not necessarily contain an element of order m.*

For example, consider the group $G = \mathbb{Z}_{20}^*$ of invertible elements in the set $\mathbb{Z}_{20}$ so that G contains

$$\phi(20) = \phi(2^2)\phi(5) = 2(4) = 8$$

elements. This group G is a commutative group of order 8. By checking the orders of each element in G, the reader will find that G does not have an element of order 8. Instead, the orders of the elements in G are all 1, 2, or 4.

*Recall that what Lagrange's Theorem says is that if a finite group G contains a subgroup H, then $|H|$ must divide $|G|$, i.e., the number of elements in the subgroup H must divide the number of elements in the group G. The theorem does **not** guarantee the existence of elements of orders that divide $|G|$.*

Remark 2.20 *In Chapter 8, we will discuss some elementary properties of error-correcting codes, where we give a beautiful application of Lagrange's Theorem to the world of modern communications.*

Before closing this chapter on groups, we mention several textbooks that discuss various topics in group theory, as well as topics in rings and fields as we will discuss in Chapters 3 and 4. These textbooks, which also discuss various topics in abstract algebra, include [4], [9], [10], [16], and [17].

2.4 Exercises

1. Prove parts (1), (2), and (4) from the list of coset properties in Theorem 2.11.

2. (a) Let $G = Z_{12}$ be the group of integers under the operation of addition modulo 12. Let H_2 denote the subset of G consisting of the integers in G that are divisible by 2. Is H_2 a subgroup of G? If so, how many cosets does it have in G? List all of the elements of the various cosets.

   (b) Answer the same questions for the subsets $H_3, H_4,$ and H_6 where H_3 denotes the set of elements of $G = Z_{12}$ that are divisible by 3. The subsets H_4 and H_6 are similarly defined.

3. Let $G = \mathbb{Z}_{16}$ be the additive group of integers modulo 16. Show that the set
$$H = \{0, 2, 4, 6, 8, 10, 12, 14\}$$
forms a subgroup of the group G. Determine the elements in each of the left cosets
$$1 + H, 2 + H, \ldots, 15 + H.$$

4. Assume that an element a in a finite commutative group G has order m. Similarly assume that an element b in G has order n, where m and n are relatively prime. Show that the element ab has order mn.

   Note that this property will be used in the proof of Theorem 5.7 where it will be shown that the multiplicative group $\mathbb{F}_q^*$ of all non-zero elements in the finite field $\mathbb{F}_q$ is cyclic.

Chapter 3

Rings

3.1   Definition of a ring

In this chapter we consider an algebraic structure that has two binary operations and builds off the idea of a group, which we already discussed.

The two operations will be thought of as addition and multiplication. Under addition, in order to form a ring, the set must form a commutative (Abelian) group. As will be seen shortly, the multiplicative operation must work in a way that is compatible with the additive operation, but the multiplicative operation itself need not be commutative.

We will take the liberty of simply writing $a \cdot b$ as ab to indicate the product of the elements a and b.

Definition 3.1 *A non-empty set R is a* **ring** *if there are two operations, addition and multiplication, for which the following properties hold:*

R1: for all $a, b \in R, a + b \in R$ (addition is closed);

R2: for all $a, b, c \in R, a + (b + c) = (a + b) + c$ (addition is associative);

R3: there is an element $0 \in R$ so that, for all $a \in R$, $a + 0 = 0 + a = a$ (identity for addition);

R4: for each $a \in R$ there is an element $-a \in R$ so that $a + (-a) = -a + a = 0$ (inverse for addition);

R5: for all $a, b \in R$, $a + b = b + a$ (addition is commutative);

R6: for all $a, b \in R$, $ab \in R$ (multiplication is closed);

R7: for all $a, b, c \in R, a(bc) = (ab)c$ (multiplication is associative);

R8: for all $a, b, c \in R, a(b + c) = ab + ac$, and $(a + b)c = ac + bc$ (multiplication distributes over addition).

Note that properties R1, R2, R3, R4, and R5 guarantee that the set R with the additive operation forms a commutative group. Property R8 is what connects the additive operation to the multiplicative operation.

We note that in a ring, the additive operation must be commutative, but the multiplicative operation need not be commutative. If the multiplicative operation in the ring is commutative, i.e., if

$$ab = ba$$

for all $a, b \in R$, then the ring R is said to be **commutative** or **Abelian**.

Further, if R contains an element 1 so that, for all $a \in R$, we have

$$a(1) = (1)a = a,$$

then the element 1 is a **unity** element for the ring R. We note that the unity element has many of the same properties in a ring as the number 1 has in the real numbers. The reader will be asked to show that a ring R can have only one unity element; see Exercise 3.1.6.

We now provide a number of examples of rings. As the reader will quickly detect, many of these examples involve objects with which we are already familiar.

1. The best example of a ring, and arguably the motivation for the various ring axioms, is the set $R = \mathbb{Z}$ of integers with the usual operations of addition and multiplication.

   Recall from our discussion on groups in Chapter 2, the integers $\mathbb{Z}$ form an Abelian group under addition. Moreover, the product of two integers is an integer, so axiom R6 is satisfied. Multiplication of real numbers is associative, so it is automatically associative on the subset of integers. Hence, R7 holds. The only axiom that really needs to be checked more carefully is the distributive axiom R8. But recall from our algebra and calculus days, this distributivity property holds for all real numbers, and hence by default, it holds for the integers. The set $\mathbb{Z}$ is thus a ring under the usual operations of addition and multiplication. The ring $\mathbb{Z}$ is actually a commutative ring with unity since multiplication of integers is commutative and the unity element in $\mathbb{Z}$ is the usual integer 1.

2. The set of all real numbers with the usual binary operations of addition and multiplication forms a commutative ring with unity.

3. The set of all rational numbers with the usual binary operations of addition and multiplication forms a commutative ring with unity.

4. The set of all $n \times n$ matrices with real entries under the usual operations of matrix addition and multiplication forms a ring. The reader can find details regarding these matrix operations in Section 6 of the Appendix. Note that this is an example of a ring that is not commutative.

5. Let R be a ring and let $R[x]$ denote the set of all polynomials in a single variable x whose coefficients are in the ring R. Define the sum

and product of two polynomials in $R[x]$ via the usual way of adding and multiplying polynomials, except that the coefficients are computed using the ring operations in R. With these definitions, $R[x]$ forms a ring, which is commutative if R itself is commutative.

6. If R is a ring, let $R[x_1, \ldots, x_n]$ denote the set of all polynomials in the n variables $x_1, \ldots, x_n$. This set also forms a ring with the usual definitions of addition and multiplication for polynomials. As in the previous example, the coefficients are computed using the operations of addition and multiplication in the ring R.

We now prove that in any ring R, $0 \cdot a = 0$ for each element $a \in R$. We calculate as follows in the ring R:

$$0 \cdot a = (0 + 0) \cdot a = 0 \cdot a + 0 \cdot a.$$

Hence, upon adding the additive inverse of the ring element $0 \cdot a$ to both sides, we see that $0 = 0 \cdot a$. Note that in the above calculation, we used the distributivity axiom, which holds for any ring.

This result may seem to be rather trivial, but it is important and states that the additive identity 0 of the ring, when multiplied by any element a in the ring, gives the additive identity 0. This result is analogous to the situation in the real numbers where we know that 0 times any real number is 0.

In the exercises we ask the reader to prove another similar kind of ring result, namely that in a ring with unity element 1, $(-1) \cdot a = -a$. This result simply says that the additive inverse -1 of the unity element 1 when multiplied in a ring by any ring element a yields the additive inverse $-a$ of the ring element a.

An element r in a ring R with unity 1 is **invertible** if there is an element $s \in R$ so that

$$rs = sr = 1.$$

An element $t \in R, t \neq 0$, is a **zero divisor** if R contains an element $u \neq 0$ so that either $tu = 0$ or $ut = 0$.

Our next result shows that in any finite ring, every non-zero element is either invertible or a zero divisor. In Exercise 3.1.7 we ask the reader to show that a similar result does not hold if the ring has an infinite number of elements.

Theorem 3.2 *If R is a finite ring with unity, then every non-zero element in R is either invertible or a zero divisor.*

Proof: Assume that a ring R has n distinct elements. Let r be a non-zero element of R. Consider the set $1 = r^0, r, r^2, \ldots, r^n$ of powers of r. This gives us a set of $n + 1$ elements. Since the ring R contains only n distinct elements, two of the powers must be equal.

Suppose that $r^i = r^{i+j}$ for some $i \geq 0, j > 0$. Then we must have

$$r^{i+j} - r^i = 0$$

so that

$$r^i(r^j - 1) = 0.$$

Now by the Well Ordering Principle (see the Appendix, Section 2 for details), there is a smallest such value of i so that

$$r^i(r^j - 1) = 0.$$

If $i = 0$, then $r^j - 1 = 0$ so that $r(r^{j-1}) = 1$. Thus, r is a unit in R (which means r is invertible in R). On the other hand, if $i > 0$, then

$$r^{i-1}(r^j - 1) \neq 0$$

by the minimality of i. However,

$$r(r^{i-1}(r^j - 1)) = r^i(r^j - 1) = 0,$$

and thus r is a zero divisor in the ring R. ∎

In some rings it is quite easy to determine if an element is invertible. For example, in the ring $\mathbb{Z}_n$ of integers modulo n, as was shown in Chapter 1, Theorem 1.33, an element a is invertible if and only if $\gcd(a, n) = 1$, i.e., if and only if the element a is relatively prime to the modulus n. This condition is quite easy to check.

In certain other rings, it may be quite tedious to determine if an element is invertible. For example, consider the ring R of all 10×10 matrices with real entries. In this setting, a matrix will be invertible if and only if it is non-singular, i.e., if and only if its determinant is not 0. This condition is easy to state, but given a random 10×10 matrix, it is not an easy task to calculate its determinant!

Definition 3.3 *The* **characteristic** *of a ring is the least positive integer n such that*

$$n \cdot 1 := \underbrace{1 + 1 + \cdots + 1}_{n \text{ copies of } 1} = 0.$$

If no such n exists, the characteristic is declared to be 0.

It is easy to see that the ring of integers has characteristic 0, as does the ring of real numbers with the usual operations of addition and multiplication.

On the other hand, the ring $\mathbb{Z}_4$ of integers modulo 4 has characteristic 4 and more generally, the ring $\mathbb{Z}_n$ has characteristic n. In the next chapter we will discuss special rings called "fields" and we will see that every field has characteristic either 0 or a prime number p.

3.1 Exercises

1. Determine which of the following form rings:

   * The set of all 3×3 upper triangular matrices with real entries under the usual operations of matrix addition and multiplication. Recall that a matrix is **upper triangular** if all of the elements that occur below the main diagonal are 0.

   * The set of all 3×3 upper triangular matrices with real entries that have ones on the main diagonal. These matrices thus have the form

   $$\begin{pmatrix} 1 & a & b \\ 0 & 1 & c \\ 0 & 0 & 1 \end{pmatrix},$$

   where $a, b,$ and c are real numbers.

   * The power set $P(X)$ (which is the set of all subsets of the set X), with union of two sets being the additive operation and intersection of two sets being the multiplicative operation.

2. Find an example of a ring R and elements $x, y \in R$ for which $(x + y)^2 \neq x^2 + 2xy + y^2$.

3. Find an example of a ring R so that, for any elements $a, b \in R$, $(a+b)^2 = a^2 + b^2$.

4. If p is a prime, find an example of a ring R so that, for any elements $a, b \in R$, $(a + b)^p = a^p + b^p$.

5. In a ring R with unity element 1, show that $(-1) \cdot a = -a$ for any ring element a.

6. If R is a ring with a unity element, show that the unity element must be unique.

7. Give an example of a ring R and an element $r \in R$ so that r is neither invertible nor a zero divisor. Hint: Recall from Theorem 3.2 that R must be a ring containing an infinite number of elements.

8. In the ring $\mathbb{Z}_{12}$ of integers modulo 12, determine which elements are invertible and which are zero divisors. Do the same problem for the ring $\mathbb{Z}_{16}$ of integers modulo 16.

9. Determine the number of invertible elements in the ring $\mathbb{Z}_{28}$ of integers modulo 28.

10. Show that in a commutative ring R, a unit cannot be a zero divisor.

11. Let R be a finite commutative ring, and let $r \in R, r \neq 0$ be an element that is not a zero divisor. If $s, t \in R$ and $rs = rt$, show that $s = t$; see Theorem 3.2.

12. Let R be a commutative ring. Show that, if $u \in R$ is a unit and $r \in R$ is **nilpotent** (so that $r^k = 0$ for some positive integer k), then $u + r$ is a unit in R. Hint: Show that the inverse of $u + r$ is given by the element

$$\frac{1}{u}\left(1 - \frac{r}{u} + \left(\frac{r}{u}\right)^2 + \cdots (-1)^k \left(\frac{r}{u}\right)^k\right)$$

where the inverse of u is denoted by $\dfrac{1}{u}$.

13. Show that the set $\{0\}$ with the usual operations of addition and multiplication is a commutative ring.

14. Let $(A, +, \cdot)$ be a ring with unity element e, where A contains at least two elements. Show that $e \neq 0$.

15. Find five distinct polynomials in $\mathbb{Z}_3[x]$ which induce the zero function on $\mathbb{Z}_3$; i.e., functions that map every element of the ring $\mathbb{Z}_3$ of integers modulo 3 to zero.

16. Let $N = \{0, 1, 2, 3\}$. Consider the tables:

+	0	1	2	3
0	0	1	2	3
1	1	0	3	2
2	2	3	0	1
3	3	2	1	0

$\cdot$	0	1	2	3
0	0	0	0	0
1	0	1	2	3
2	0	1	2	3
3	0	0	0	0

Show that N is a non-commutative ring. Show that N has zero divisors and that N does not have a unity element.

17. Let R be a ring such that the additive group $\{R, +\}$ is a cyclic group. Show that R is a commutative ring.

18. Given two rings $(A, +, *)$ and $(B, +, \cdot)$, we can form a third ring using the Cartesian product $A \times B$ of the sets A and B. In particular, for $a, a' \in A$ and $b, b' \in B$, we define

$$(a, b) + (a', b') = (a + a', b + b').$$

Multiplication is similarly defined. Verify that with these operations, the Cartesian product $A \times B$ is indeed a ring.

3.2   Subrings and ideals

As in the case of groups, some rings have the property that some of their substructures satisfy very nice conditions. We begin with a definition that mimics the concept of a subgroup.

Definition 3.4 *Let R be a ring. A subset B of R is a* **subring** *if B itself forms a ring, with the same operations of addition and multiplication that are used in R.*

In addition, a subring I of a ring R is an **ideal** *of R if the product of any element in R and any element in I is always an element of I. Thus the subring I is an ideal if for each $i \in I$ and for each $a \in R$, we have that $ia \in I$ and $ai \in I$.*

We note that, in particular, for a subset B of a ring R to be a subring, the sum and product of two elements of B must be in B. The condition to be satisfied if B is to be an ideal is even stronger; namely B must be closed under multiplication if one of the two elements is in R but is outside of the subset B.

The set I of all multiples of 3 (or of any fixed integer) is a subring of the ring $\mathbb{Z}$ of integers. In fact, the set I forms an ideal in the ring $\mathbb{Z}$ of integers since, if $n \in \mathbb{Z}$ and $3k \in I$, then

$$n(3k) = 3(nk) \in I.$$

A similar example of an ideal can be found by letting the ring $R = \mathbb{Z}_6$ be the ring of integers modulo 6 and taking the subset B to be $B = \{0, 2, 4\}$.

We now give several examples of subrings that do not form ideals. Consider the ring $\mathbb{Z}$ integers, which forms a subring of the rational numbers $\mathbb{Q}$. Even so, note that $\mathbb{Z}$ is not an ideal in $\mathbb{Q}$ because the product of an integer and a rational number can be a rational number that is not in $\mathbb{Z}$. For example, $5 \in \mathbb{Z}$ and $1/7 \in \mathbb{Q}$ but $5/7 \notin \mathbb{Z}$. Also, $\mathbb{Q}$ is a subring but not an ideal in the ring of real numbers.

Another example can be obtained by considering the set of all 2×2 matrices of the form

$$\begin{pmatrix} m & 0 \\ 0 & 0 \end{pmatrix}$$

where m is an integer. These matrices form a subring, but not an ideal, in the ring of all 2×2 matrices over the integers. The reader should verify that this is indeed the case.

The reader should also check that the set I of all 2×2 matrices with even integer entries is a subring, in fact, an ideal, in the ring of all 2×2 matrices with integer entries. The key fact to verify that I is an ideal follows from the matrix calculation

$$\begin{pmatrix} 2a & 2b \\ 2c & 2d \end{pmatrix} \begin{pmatrix} e & f \\ g & h \end{pmatrix} = \begin{pmatrix} 2ae + 2bg & 2af + 2bh \\ 2ac + 2dg & 2cf + 2dh \end{pmatrix} \in I,$$

since each entry is an even integer. Thus I is an ideal in the ring of all 2×2 matrices with integer entries.

The reader should now prove the following two theorems (see Exercises 3.2.4 and 3.2.5).

Theorem 3.5 *A non-empty set B of a ring R is a subring of R if and only if $a - b$ and ab are in B for all $a, b \in B$.*

Theorem 3.6 *A non-empty subset I of a ring R is an ideal in R if $a - b \in I$ for all $a, b \in I$ and xa and ax are in I for every $x \in I$ and every $a \in R$.*

Definition 3.7 *Let I be an ideal in a ring R and let $a \in R$. Then the set*

$$a + I = \{a + k | k \in I\}$$

*is a **left coset** determined by a and I. The set R/I of all left cosets with operations defined by*

$$(a + I) + (b + I) = (a + b) + I$$

and

$$(a + I)(b + I) = (ab) + I$$

*is the **quotient ring** of R modulo the ideal I.*

In order to justify use of the word "ring" with respect to R/I, it is first necessary to prove the following theorem.

Theorem 3.8 *Let I be an ideal in a ring R. Then $\{R/I, +, \cdot\}$ is a ring.*

For the proof, see Exercises 3.2.10 and 3.2.11.

3.2 Exercises

1. Show that $\mathbb{Q}$ is a subring of the ring of real numbers $\mathbb{R}$ but that $\mathbb{Q}$ is not an ideal of $\mathbb{R}$.

2. Find all ideals in the ring $\mathbb{Z}_8$ of integers modulo 8.

3. Show that $\{0\}$ and R are always ideals in any ring R.

4. Prove Theorem 3.5 for subrings.

5. Prove Theorem 3.6 for ideals.

6. Show that the set of all 2×2 matrices with even entries is an ideal in the ring of 2×2 matrices with integer entries.

7. If I and J are ideals in a ring R, show that $I \cap J$ is an ideal of R. Is it true that $I \cup J$ must also be an ideal of R?

8. Extend the result of the previous exercise to show that the intersection of any finite number of ideals in a ring is an ideal.

9. Assume that I and J are ideals in a ring R. Show that the set

$$I + J = \{i + j | i \in I, j \in J\}$$

is an ideal in the ring R.

10. Show that multiplication in R/I is well–defined by showing that, if $x \in a + I$ and $y \in b + I$, then $xy \in ab + I$.

11. Prove Theorem 3.8.

12. Let R be a commutative ring and let $r \in R$. Show that the set $P_r = \{xr | x \in R\}$ is an ideal in R. The ideal P_r is the **principal ideal** generated by the element r.

13. Let R be a ring with a unity element e. Let I be an ideal of R with $e \in I$. Show that $I = R$.

3.3   Ring homomorphisms

We begin with a definition that will lead to a description of what is meant by saying that two rings are the "same."

Definition 3.9 *Assume that $\{A, +, *\}$ and $\{B, \oplus, \cdot\}$ are rings. A **ring homomorphism** is a mapping $f : A \to B$ such that*

$$f(a + b) = f(a) \oplus f(b) \text{ and } f(a * b) = f(a) \cdot f(b)$$

*for all $a, b \in A$. If f is both 1-1 and onto, then f is an **isomorphism** and the rings A and B are **isomorphic**.*

Thus a mapping between two rings is a homomorphism if it preserves both the additive and the multiplicative operations. Moreover, if A and B are isomorphic, then in some sense they are basically the "same."

We now give a few examples of ring homomorphisms.

Example 3.10 *For any ring R, define $f : R \to R$ by $f(r) = r$ for each r in R. It is easy to check that this function f is indeed a ring homomorphism. (Indeed, it is a ring isomorphism, as the reader should verify.)*

Example 3.11 *Consider the ring $\mathbb{Z}$ of integers with the usual operations of addition and multiplication. Define a function $f : \mathbb{Z} \to \mathbb{Z}_4$ that assigns to each integer a the least non-negative residue of a modulo 4. Is this mapping a homomorphism onto the ring $\mathbb{Z}_4$? Is it $1-1$?*

We first verify that this mapping is a ring homomorphism. To this end, by the Division Algorithm, we may write $a = 4k + r$ for some integer k and where $0 \le r < 4$. Similarly, we can assume that $b = 4l + s$ for some integer l with $0 \le s < 4$. Then

$$a + b = 4(k + l) + (r + s).$$

Hence the least non-negative residue of $a + b$ is $r + s$, which is the sum modulo 4 of r, the least non-negative residue of a plus s, the least non-negative residue of b. Hence we see that $f(a + b) = f(a) + f(b)$.

The reader should show via a similar calculation that $f(ab) = f(a)f(b)$.

Is this mapping onto? Sure, since if $a \in \mathbb{Z}_4$ so that $a = 0, 1, 2$ or 3, then $f(a) = a$, and thus the mapping is onto. Is it $1-1$? No, since $f(2) = f(6) \equiv 2$ (mod 4).

3.3 Exercises

1. Let f be a homomorphism from a ring A onto a ring B. Let B_1 be a subring (ideal) of B. Show that

   $$f^{-1}(B_1) = \{x \in A | f(x) \in B_1\}$$

   is a subring (ideal) of A.

2. Define a homomorphism from the ring $\mathbb{Z}$ of integers to the ring $\mathbb{Z}_6$ of integers modulo 6.

3. Define the product MN of subsets of a ring R by

   $$MN = \{mn | m \in M, n \in N\}.$$

   Let I be an ideal in R. Show that

   $$(a + I)(b + I) \subset ab + I.$$

4. If $f : A \to B$ is a ring homomorphism, show that the image of f is a subring of B. Recall from Section 4 of the Appendix that the image of a function $f : A \to B$ is the set $\{f(a) | a \in A\}$.

5. If $f : A \to B$ is a ring homomorphism, show that the set

   $$K = \{a \in A | f(a) = 0\}$$

   is an ideal in A. K is called the **kernel** of the ring homomorphism f.

3.4 Integral domains

Definition 3.12 *An* **integral domain** *is a commutative ring with unity, which does not contain any zero divisors.*

If one is studying integral domain properties from other books, one must be careful as some authors omit some of the conditions of commutativity, unity, or both when discussing integral domains.

The best example of an integral domain is the ring $\mathbb{Z}$ of integers with the usual operations of addition and multiplication.

One of the interesting properties of integral domains is a kind of cancellation property (because of the lack of zero divisors). We now prove this result.

Theorem 3.13 *Let R be an integral domain with $a, b, c \in R, a \neq 0$. If $ab = ac$, then $b = c$.*

Proof: If $ab = ac$ then $ab - ac = a(b - c) = 0$. The ring R is an integral domain and thus does not have any zero divisors. Since $a \neq 0$, then $b - c = 0$ and hence $b = c$. ∎

We note that a converse of this theorem also holds.

Theorem 3.14 *If R is a commutative ring with unity, with the property that for any non-zero $a \in R$, if $ab = ac$ implies that $b = c$, then R is an integral domain.*

Proof: Let $a \in R$ be a non-zero element. Assume that $ab = 0$ so that $ab = a(0)$. Then $b = 0$, so the ring is an integral domain. ∎

Definition 3.15 *Let D be an integral domain and $a, b \in D$. Then a **divides** b (a is a **factor** of b or b is a **multiple** of a) if there is an element $c \in D$ so that $ac = b$. If a divides b, we write $a|b$ and if a does not divide b, we write $a \nmid b$. Elements a and b in an integral domain are called **associates** if a divides b and b divides a. An associate of the unity element e is a **unit**.*

We ask the reader to prove the following results in Exercise 3.4.4.

Theorem 3.16 *Let D be an integral domain and let $a, b, c, x \in D$. Then the following hold:*

(i) If $a|b$ and $b|c$, then $a|c$;

(ii) An element x is a unit of D if and only if x^{-1} exists in D;

(iii) An element x is a unit of D if and only if x divides every element of D;

(iv) Elements a and b are associates if and only if there exists a unit $x \in D$ such that $ax = b$.

Before closing this chapter on rings, we refer to several textbooks dealing with abstract algebra, which provide further discussion on various properties of rings. These textbooks include [4], [9], [10], [16], and [17].

3.4 Exercises

1. Consider the ring N as defined in Exercise 3.1.16. Find a subring of N that is an ideal.

2. Let $C[0,1]$ denote the set of all continuous functions on the interval $[0,1]$. Thus $C[0,1]$ consists of all functions

$$f : [0,1] \to [0,1]$$

   that are continuous.

   Show that $C[0,1]$ is a ring with respect to the usual operations of addition and multiplication of functions. Also, show that $C[0,1]$ is not an integral domain.

3. Prove that if R is an integral domain, then any subring of R that contains a unity element is an integral domain.

4. Prove each of the four parts of Theorem 3.16.

5. Let R be the ring $\mathbb{Z}_8$ of integers modulo 8. Show that the set $S = \{0,2,4,6\}$ is a subring of the ring R. Is the set S an ideal in the ring R?

6. Show that the product of two units in a ring is also a unit. Note that this result extends Theorem 1.35 for the group $\mathbb{Z}_n^*$ of invertible elements in $\mathbb{Z}_n$, to a general ring R.

7. For which values of n is $\mathbb{Z}_n$ an integral domain?

Chapter 4

Fields

4.1   Definition and basic properties of a field

In this chapter we again consider an algebraic structure that has two binary operations. The two operations will be addition and multiplication. We impose an extra condition on the non-zero elements in the set and arrive at the following definition:

Definition 4.1 *A **field** is a commutative ring with a unity element with the additional property that each non-zero element has a multiplicative inverse, i.e., each non-zero element in the field is a unit.*

We note that a field is a commutative ring in which the set of all non-zero elements is a group under the operation of multiplication. We already know some examples of fields; we mention the following for which the reader should carefully verify that the necessary properties hold.

1. The set $\mathbb{R}$ of all real numbers with the usual operations of addition and multiplication.

2. The set

$$\mathbb{Q} = \left\{ \frac{a}{b} \mid a, b \in \mathbb{Z}, b \neq 0 \right\}$$

   of all rational numbers with the usual operations of addition and multiplication of fractions. Thus we define

$$\frac{a}{b} + \frac{c}{d} = \frac{ad + bc}{bd}$$

and

$$\frac{a}{b}\frac{c}{d} = \frac{ac}{bd}.$$

3. The set $\mathbb{C}$ of all complex numbers where we define addition and multiplication by

$$(a + bi) + (c + di) = (a + c) + (b + d)i$$

and

$$(a + bi)(c + di) = (ac - bd) + (ad + bc)i,$$

recalling that $i = \sqrt{-1}$ so that $i^2 = -1$.

See Section 7 of the Appendix for further discussion of complex numbers.

Examples 1–3 are **infinite** fields, i.e., fields that contain an infinite number of elements. Are there fields that contain only a finite number of elements? The next example of a field shows that the answer is yes, there are fields that contain only a finite number of elements.

4. If p is a prime, then the commutative ring $\mathbb{Z}_p$ of integers modulo p is a field with p elements.

For any prime p, the reader should check that the necessary field axioms are satisfied for the set $\mathbb{Z}_p$ of integers modulo p. Thus there is an infinite number of finite fields, one for each of the infinitely many primes p.

Finite fields will be discussed in much further detail in Chapter 5, where it will be shown that finite fields exist that contain a finite, but not necessarily prime, number of elements.

Consider the set $\mathbb{Z}[\sqrt{2}]$ of all real numbers of the form $a + b\sqrt{2}$, where a and b are integers. The reader should check that with the usual operations of addition and multiplication as in the reals, the set $\mathbb{Z}[\sqrt{2}]$ is a commutative ring. The reader should also check that this ring does not form a field. Why?

Now consider the larger set $\mathbb{Q}[\sqrt{2}]$ consisting of all real numbers of the form $a + b\sqrt{2}$ where a and b are now allowed to be rational numbers. Does this larger set form a field? It does, as the reader should carefully check. The most difficult point to check is that every non-zero element indeed has a multiplicative inverse. To this end, assume that $a + b\sqrt{2}$ is non-zero (so that at least one of a or b is not 0) and hence $a^2 - 2b^2 \neq 0$. The reader should check that

$$(a + b\sqrt{2}) \times (c + d\sqrt{2}) = 1$$

holds if $c = a/(a^2 - 2b^2)$ and $d = -b/(a^2 - 2b^2)$. Hence the element $c + d\sqrt{2}$ is indeed the inverse of the element $a + b\sqrt{2}$.

We now illustrate how to construct a field from any integral domain. Recall that an integral domain is a commutative ring with unity which does not have any zero divisors; i.e., no non-zero element a so that $ab = 0$ for any non-zero element b in the ring.

In fact, in the case of the integral domain $\mathbb{Z}$ of integers, this process yields the usual field $\mathbb{Q}$ of rational numbers! In general, the field resulting from the following construction is called the **field of fractions**, or the **quotient field**.

Let I be an integral domain. Consider the set X of all pairs (r, s) of elements of I with the element $s \neq 0$. On the set X, we now define two pairs (r, s) and (t, u) to be related, i.e., $(r, s)R(t, u)$, if and only if $ru = st$. We note that in the case of fractions $\frac{r}{s}$ and $\frac{t}{u}$ this relation is exactly what we use to say that the two fractions are equal as rational numbers. The reader should show in Exercise 4.1.23 that this relation R is an equivalence relation.

Let $\mathbb{Q}$ be the set of all equivalence classes $[(r, s)]$ of the relation R; see the Appendix, Section 3, for a discussion of equivalence classes arising from an equivalence relation. We now define two operations of addition and multiplication on $\mathbb{Q}$ by defining

$$[(r, s)] + [(t, u)] = [(ru + st, su)]$$

and

$$[(r, s)][(t, u)] = [(rt, su)].$$

In Exercise 4.1.22 the reader will be asked to show that these operations are indeed well defined; i.e., that they do not depend on the representatives taken from the equivalence classes.

It is not too hard to show that $\mathbb{Q}$ is a commutative ring with these operations. Moreover, every non-zero element of $\mathbb{Q}$ has an inverse and thus $\mathbb{Q}$ is a field.

We now define a function $f : I \to \mathbb{Q}$ by mapping an element $r \in I$ to the equivalence class of $(r, 1)$, i.e., to the equivalence class $[(r, 1)]$. The reader should show that this function is both $1 - 1$ and onto.

In addition, we have that $f(r + t) = f(r) + f(t)$ and $f(rt) = f(r) \times f(t)$. This shows that there is a copy of the integral domain $\mathbb{Z}$ embedded inside the field $\mathbb{Q}$. Finally, once one shows that every element of $\mathbb{Q}$ has the form $f(r) \times f(s)^{-1}$ for some $r, s \in I$ with $s \neq 0$, the field $\mathbb{Q}$ thus consists of all fractions formed from elements of I.

This justifies calling the field $\mathbb{Q}$ the **field of fractions** or the **quotient field** of the integral domain I.

Moreover, if the integral domain $I = \mathbb{Z}$ the ring of integers, then the field $\mathbb{Q}$ is simply the ordinary field of rational numbers discussed in Example 2 at the beginning of this chapter.

In Exercise 4.1.1 we ask the reader to show that a field is an integral domain; i.e., a field cannot contain any zero divisors. The following result shows that the converse holds if the integral domain is finite. In Exercise 4.1.19 we ask the reader to provide an example of an integral domain (which must be infinite) that is not a field.

Theorem 4.2 *A finite integral domain is a field.*

Proof: The only property that must be verified is the existence of multiplicative inverses. Let R be a finite integral domain and choose a non-zero element $a \in R$. Consider the set

$$aR = \{ar | r \in R\}.$$

If $r_1 \neq r_2$, we must have $ar_1 \neq ar_2$, for if $ar_1 = ar_2$, then $a(r_1 - r_2) = 0$. Since $a \neq 0$, we would have $r_1 - r_2 = 0$ and hence $r_1 = r_2$, for otherwise the ring would contain zero divisors. Hence $|aR| = |R|$, so the unity element $1 \in aR$. Thus there is an element $b \in R$ such that $ab = 1$, and b is thus the inverse of the element a. ∎

We close this chapter by considering the notion of a subfield as given in the next definition.

Definition 4.3 *A non-empty subset S of a field F is a* **subfield** *if S is itself a field under the same operations as in the field F.*

Examples of subfields are easy to locate. The rational numbers form a subfield of the field of real numbers and the real numbers themselves form a subfield of the field of complex numbers, as the reader should check. Similarly, the rational numbers form a subfield of the field of complex numbers.

In the next chapter we will discuss finite fields and we will see that finite fields can have many subfields; in fact, we can easily enumerate and construct them all.

Before closing the discussion in this chapter on fields, we refer to several textbooks dealing with abstract algebra, which provide further discussion of various properties of fields. These textbooks include [4], [9], [10], [16], and [17].

4.1 Exercises

1. Show that a field cannot have any zero divisors. Thus show that if F is a field and $a, b \in F$ satisfy $ab = 0$, then either $a = 0$ or $b = 0$.

   We remind the reader that this very basic property, which holds for any field, is a property that we have used hundreds of times when solving equations over the real numbers. When trying to obtain the roots of a polynomial equation, we usually first factor the polynomial into a product of factors, and then we set each of the factors equal to zero to obtain the roots of the equation.

2. Let R be a ring and let S be the intersection of all subrings of R. Show that S is a subring of R, and that if R is a field, then S is also a field.

3. Let F be a field. Show that the polynomial ring $F[x]$ consisting of all polynomials whose coefficients are in the field F has no zero divisors.

4. Prove that in a field F, if $a \neq 0$ and $b \in F$, then the equation $ax = b$ has a unique solution in the field F.

5. Show that the ring $\mathbb{Z}_n$ of integers modulo n is a field if and only if n is a prime number.

6. Does the equation $x^2 + 5 = 0$ have a solution in the field of real numbers?

7. Does the equation $x^2 + 5 = 0$ have a solution in the field of complex numbers? If so, find all of the solutions.

8. Explain why the commutative ring $\mathbb{Z}_8$ of integers modulo 8 does not form a field or an integral domain.

9. Show that if F is a field and a is any element in F, then $a \cdot 0 = 0$. This result also holds for any ring.

10. For any prime $p \neq 2$, show that in the field $\mathbb{Z}_p$ of integers modulo p, the sum of the elements in the field $\mathbb{Z}_p$ is 0.

11. Factor the polynomial $x^4 + 1$ over the field $\mathbb{Z}_2$ of integers modulo 2.

12. List all polynomials of degree at most three whose coefficients lie in the field $\mathbb{Z}_3$ of integers modulo 3.

13. In the field $\mathbb{Z}_7$ of integers modulo 7, find the multiplicative order of each non-zero element. Recall that the multiplicative order of an element a in the ring $\mathbb{Z}_n$ is the least exponent k so that $a^k \equiv 1 \pmod{n}$ in the ring.

14. Assume that F is a field and a is a non-zero element of F. Show over the field F that if $ax = ay$, then $x = y$.

15. Prove: If $-a = a^{-1}$ for all non-zero elements $a \in F$ a field, then $F = \mathbb{Z}_2$. (Hint: First show that for such a, $-a = a$. Then show $a^2 = 1$ so that $(a+1)^2 = 0$ and that $a = 1$ and thus $F = \{0, 1\}$. Hence $F = \mathbb{Z}_2$, the ring of integers modulo 2.)

16. Show that the characteristic of a field must either be a prime or 0. Recall that the characteristic of a ring is the smallest integer k so that the sum $1 + \cdots + 1$, k times, equals 0. If no such positive k exists, then the characteristic is defined to be 0.

17. Show that if F is a field of characteristic 2, then $-a = a$ for any $a \in F$.

18. If a field F has characteristic p a prime, show for any $a, b \in F$ that $(a+b)^p = a^p + b^p$.

19. Give an example of an integral domain (which by Theorem 4.1 must be infinite) that is not a field.

20. If a field F has characteristic p a prime, show for any $a, b \in F$ that $(a+b)^{p^n} = a^{p^n} + b^{p^n}$ for any positive integer n.

21. Consider the set F consisting of the following four 2×2 matrices with entries from $\mathbb{Z}_2$, the ring of integers modulo 2:

$$\begin{pmatrix} 0 & 0 \\ 0 & 0 \end{pmatrix}, \begin{pmatrix} 1 & 0 \\ 0 & 1 \end{pmatrix}, \begin{pmatrix} 1 & 1 \\ 1 & 0 \end{pmatrix}, \begin{pmatrix} 0 & 1 \\ 1 & 1 \end{pmatrix}.$$

Show that F is a field. What is the characteristic of the field F?

22. Show that the operations of addition and multiplication defined in the field of fractions construction in Section 4.1 are indeed well defined; i.e., that they do not depend on the representatives taken from the equivalence classes.

23. Consider the set X of all pairs (r, s) of elements of an integral domain I with the element $s \neq 0$. On the set X, define two pairs (r, s) and (t, u) to be related, i.e., $(r, s)R(t, u)$ if and only if $ru = st$. Show that this relation R forms an equivalence relation, i.e., R is reflexive, symmetric, and transitive.

24. If E is a subfield of a field F, is the set $E[x]$ of all polynomials with coefficients in E a subfield of the ring $F[x]$? Is it a subring of $F[x]$?

25. Show that the intersection of two subfields of a field F is a subfield of F. Is it true that the intersection of any finite collection of subfields of a given field F is a subfield of F?

Chapter 5

Finite Fields

We know from the previous chapter that the real numbers, the rational numbers, and the complex numbers form fields (under the usual operations of addition and multiplication). Each of these fields contains an infinite number of distinct elements.

This raises an obvious question: Are there **finite sets** that also form fields? If so, how many elements do such fields contain? How can they be constructed? Are they useful? We will answer each of these questions in this chapter.

5.1   Number of elements in a finite field

The answer to the first motivating question above is a resounding "Yes!" Thus, we begin our study of finite fields by determining the possible sizes or cardinalities of a finite field. For example, given a positive integer $m \geq 2$, is there a finite field with exactly m distinct elements? If so, how can one construct such a field?

As a first step, we already know from Chapter 1 that, for any prime p, there is a finite field with exactly p elements, namely the set

$$\{0, 1, \ldots, p-1\} = \mathbb{Z}_p$$

of integers modulo the prime p. Since there are infinitely many primes, this proves that there are infinitely many finite fields.

Now consider a positive integer $m \geq 2$ that is not a prime. Is there a finite field with m distinct elements? As examples, we will soon see that there is a finite field with $m = 4, 8, 9$, and 16 elements, respectively. Notice the special property of these integers:

$$4 = 2^2, 8 = 2^3, 9 = 3^2, 16 = 2^4.$$

Each is a power of a prime number. That is, each of these numbers contains only one prime in its prime factorization.

Also notice that any prime is also of this form since any prime p can be written as $p = p^1$. Integers of the form p^n with p a prime are called **prime powers**.

It turns out that the prime powers are precisely the integers m for which there is a finite field with exactly m elements. We state this more formally as follows:

Theorem 5.1 *For any prime p and any positive integer n there is a finite field with p^n elements. Moreover, any finite field must have a prime power number of elements.*

If a finite field has p^n elements, we say that the field has **order** p^n.

We can conclude from the above theorem that, for example, there are no finite fields of orders $6, 10, 12, 14$, or 18. On the other hand, there are finite fields that contain $4, 8, 9$, or 16 elements.

How does one prove a result such as Theorem 5.1? One approach is to use the theory of vector spaces, which will be discussed in Chapter 6, to prove that a finite field must have a prime power number of elements.

To completely prove this result in detail, while not beyond the scope of our text, would take us too far out of our way. Thus, instead of proving the result, we will focus on the actual construction of finite fields.

Before constructing some finite fields with a non-prime number of elements, we point out from Theorem 5.1 that we have two infinite sequences of finite fields. For example, for a fixed prime p we can vary the value of n to be any positive integer, so that we have a field with p^n elements. In addition, for a fixed positive integer n, we can let the prime p run through the infinite set of primes to again construct an infinite number of finite fields with p^n elements.

5.1 Exercises

1. Is there a field with 3125 distinct elements? If so, what is the characteristic of the field? We remind the reader (from Definition 3.3) that the characteristic n of a ring with unity is the smallest positive integer n, so that when adding $1 + \cdots + 1$, n times, we obtain the 0 of the ring. In the case of a finite field, the characteristic must be a prime.

2. Is there a field of order 288?

3. Is there a field of order 729?

4. Is there a field of order 10046?

5.2 How to construct finite fields

If p is a prime and $n \geq 2$ is a positive integer, how do we construct a finite field with p^n elements? We begin by first considering irreducible polynomials over the field $\mathbb{Z}_p$ of p elements. (These irreducible polynomials are key ingredients in constructing finite fields.)

Let $N_p(n)$ be the number of **monic** (coefficient of the highest power of x is 1) irreducible polynomials of degree n over the field $\mathbb{Z}_p$ of integers modulo the prime p.

Recall that an irreducible polynomial is one that cannot be factored into a product of two polynomials each of smaller degree.

The following formula determines the number of monic irreducible polynomials of degree n over the integers modulo a prime p. This result was first obtained by Carl Friedrich Gauss (1777 – 1855), one of the most prolific mathematicians of the early nineteenth century.

Theorem 5.2 *For any prime p and any positive integer n, the number $N_p(n)$ of monic irreducible polynomials of degree n over the field $\mathbb{Z}_p$ is given by*

$$N_p(n) = \frac{1}{n} \sum_{d|n} \mu(d) p^{n/d}, \tag{5.1}$$

where μ is the Möbius function.

The Möbius function is defined on the positive integers by the rule

$$\mu(m) = \begin{cases} 1 & \text{if } m = 1, \\ (-1)^k & \text{if } m = m_1 m_2 \cdots m_k, \text{ where the } m_i \text{ are distinct primes,} \\ 0 & \text{otherwise; i.e., if } p^2 \text{ divides } m \text{ for some prime } p. \end{cases}$$

If we study Equation (5.1) we can show that it implies that $N_p(n) > 1$ for all primes p and all integers $n > 1$. This follows from the following inequality:

$$N_p(n) = \frac{1}{n} \sum_{d|n} \mu(d) p^{n/d} \geq \frac{1}{n}(p^n - p^{n-1} - p^{n-2} - \cdots - p) > 0.$$

Since $N_p(n)$ is positive, it must be at least one so that for any prime p and any positive integer n, we must have at least one monic irreducible polynomial of degree n over $\mathbb{Z}_p$.

Theorem 5.3 *For any prime p and any positive integer $n \geq 1$ there is an irreducible polynomial of degree n over $\mathbb{Z}_p$.*

The more difficult part of the story is how to find or construct at least one such irreducible polynomial even if we know that one exists!

In [13] can be found a list of irreducible (actually primitive) polynomials of degree n over $\mathbb{Z}_p$ for each prime $p \leq 97$ with $p^n \leq 10^{50}$. A **primitive polynomial** is a polynomial each of whose roots is a primitive element in the field $\mathbb{F}_{p^n}$. A primitive polynomial is always irreducible, so it cannot be factored into a product of polynomials of smaller degrees. We will discuss primitive elements in finite fields a bit later.

We recall (from Chapter 1) the notion of primitive elements in $\mathbb{Z}_p$ being elements of maximal multiplicative order $p - 1$. Primitive elements in general finite fields are also elements of maximal multiplicative order. In particular, in the finite field $\mathbb{F}_q$, a **primitive element** is simply an element of order $q - 1$.

We now illustrate how to construct finite fields containing 2^2 and 3^2 elements. At this point, the reader should carefully work through and try to understand all of the various calculations. Motivation for these calculations will come shortly.

Example 5.4 *Consider the polynomial $p(x) = x^2 + x + 1$ over the field $\mathbb{Z}_2$ of integers modulo 2. Since $p(x)$ does not have a root in $\mathbb{Z}_2$ (recall that $p(x)$ will have a root $a \in \mathbb{Z}_2$ if and only if $p(a) = 0$ for some $a \in \mathbb{Z}_2$ where the arithmetic is computed modulo 2), $p(x)$ is irreducible over $\mathbb{Z}_2$.*

Let θ be a root of $p(x)$ in some field larger than $\mathbb{Z}_2$ so that

$$\theta^2 + \theta + 1 = 0,$$

that is,

$$\theta^2 = -(\theta + 1) = \theta + 1$$

since we are computing modulo 2.

Actually, since the degree of the polynomial is 2, a root θ can always be found in the field $\mathbb{F}_{2^2}$, i.e., in the field containing four distinct elements.

A similar situation often occurs in the rational and real numbers. For example, the polynomial $x^2 - 5$ does not have a root in the field $\mathbb{Q}$ of rational numbers, but it does have roots $\pm\sqrt{5}$ in the larger field of real numbers.

We now give the addition and multiplication tables for the field $\mathbb{F}_{2^2}$.

$+$	0	1	θ	$\theta + 1$
0	0	1	θ	$\theta + 1$
1	1	0	$\theta + 1$	θ
θ	θ	$\theta + 1$	0	1
$\theta + 1$	$\theta + 1$	θ	1	0

×	0	1	θ	$\theta+1$
0	0	0	0	0
1	0	1	θ	$\theta+1$
θ	0	θ	$\theta+1$	1
$\theta+1$	0	$\theta+1$	1	θ

We note that

$$\theta(\theta+1) = \theta^2 + \theta = (\theta+1) + \theta = 2\theta + 1 = 1.$$

We also note that θ has the property that $\theta, \theta^2, \theta^3$ give each of the three non-zero elements of the finite field $\mathbb{F}_{2^2}$. Note that

$$\theta^1 = \theta, \theta^2 = \theta + 1, \theta^3 = 1.$$

The element θ is an example of a primitive element that lies in the field $\mathbb{F}_{2^2}$ of four elements.

In addition, the polynomial $p(x)$ is a primitive polynomial since its roots θ and $\theta+1$ are both primitive elements.

We note that in each row and each column of the addition table, we have each of the four field elements exactly once. A similar situation occurs in the multiplicative table except for the first row and first column, which only consist of the 0 element.

We also note that to locate the multiplicative inverse of a nonzero element, we go across that row until we locate the identity element 1. The element in the border at the top of that column is the inverse. Thus the multiplicative inverse of the element θ is seen to be the element $\theta + 1$.

We now consider a larger example of a finite field with $3^2 = 9$ elements. Again, motivation for these calculations will be discussed shortly.

Example 5.5 *We construct the field* $\mathbb{F}_9 = \mathbb{F}_{3^2}$ *containing 9 elements. Consider the polynomial* $p(x) = x^2 + x + 2$ *over the field* $\mathbb{Z}_3$ *of integers modulo 3. This polynomial has no roots in* $\mathbb{Z}_3$, *so it is irreducible over* $\mathbb{Z}_3$.

To show that the quadratic polynomial $p(x)$ *has no roots in the field* $\mathbb{Z}_3$ *of integers modulo 3, we simply calculate* $p(0) = 2, p(1) = 4 = 1$, *and* $p(2) = 8 = 2$ *when calculating modulo 3 in the field* $\mathbb{Z}_3$. *Since none of the three values is zero, the polynomial does not have a root in the field. Hence the polynomial is irreducible over the field* $\mathbb{Z}_3$.

Let θ *be a root of* $p(x)$, *so*

$$\theta^2 + \theta + 2 = 0.$$

Hence

$$\theta^2 = -\theta - 2 = 2\theta + 1$$

(recall we are working modulo 3).

A polynomial of degree three will be irreducible if and only if it does not have a root in the field. Check this!

Be careful, however, when checking whether a polynomial of degree, say, $n > 3$ is irreducible; it is not enough to just check whether it has a root.

For example, the polynomial $(x^2 + x + 1)^2$ is clearly reducible over the field $\mathbb{Z}_2$ since it is the square of a polynomial over the field $\mathbb{Z}_2$ of integers modulo 2. Clearly, however, neither 0 nor 1 are roots, so it does not have roots in the field $\mathbb{Z}_2$.

The field $\mathbb{F}_{3^2}$ can be represented as the set $\{a\theta + b \mid a, b \in \mathbb{Z}_3\}$ with operations being performed modulo 3. We can compute the addition and multiplication tables by hand. For example,

$$2\theta(\theta + 2) = 2\theta^2 + 4\theta = 2(2\theta + 1) + \theta = 2\theta + 2.$$

With some effort, the following addition and multiplication tables are obtained.

+	0	1	2	θ	$\theta+1$	$\theta+2$	2θ	$2\theta+1$	$2\theta+2$
0	0	1	2	θ	$\theta+1$	$\theta+2$	2θ	$2\theta+1$	$2\theta+2$
1	1	2	0	$\theta+1$	$\theta+2$	θ	$2\theta+1$	$2\theta+2$	2θ
2	2	0	1	$\theta+2$	θ	$\theta+1$	$2\theta+2$	2θ	$2\theta+1$
θ	θ	$\theta+1$	$\theta+2$	2θ	$2\theta+1$	$2\theta+2$	0	1	2
$\theta+1$	$\theta+1$	$\theta+2$	θ	$2\theta+1$	$2\theta+2$	2θ	1	2	0
$\theta+2$	$\theta+2$	θ	$\theta+1$	$2\theta+2$	2θ	$2\theta+1$	2	0	1
2θ	2θ	$2\theta+1$	$2\theta+2$	0	1	2	θ	$\theta+1$	$\theta+2$
$2\theta+1$	$2\theta+1$	$2\theta+2$	2θ	1	2	0	$\theta+1$	$\theta+2$	θ
$2\theta+2$	$2\theta+2$	2θ	$2\theta+1$	2	0	1	$\theta+2$	θ	$\theta+1$

$\times$	0	1	2	θ	$\theta+1$	$\theta+2$	2θ	$2\theta+1$	$2\theta+2$
0	0	0	0	0	0	0	0	0	0
1	0	1	2	θ	$\theta+1$	$\theta+2$	2θ	$2\theta+1$	$2\theta+2$
2	0	2	1	2θ	$2\theta+2$	$2\theta+1$	θ	$\theta+2$	$\theta+1$
θ	0	θ	2θ	$2\theta+1$	1	$\theta+1$	$\theta+2$	$2\theta+2$	2
$\theta+1$	0	$\theta+1$	$2\theta+2$	1	$\theta+2$	2θ	2	θ	$2\theta+1$
$\theta+2$	0	$\theta+2$	$2\theta+1$	$\theta+1$	2θ	2	$2\theta+2$	1	θ
2θ	0	2θ	θ	$\theta+2$	2	$2\theta+2$	$2\theta+1$	$\theta+1$	1
$2\theta+1$	0	$2\theta+1$	$\theta+2$	$2\theta+2$	θ	1	$\theta+1$	2	2θ
$2\theta+2$	0	$2\theta+2$	$\theta+1$	2	$2\theta+1$	θ	1	2θ	$\theta+2$

As in the earlier example of a field with four elements, we note that in this larger field, each nonzero element indeed has a multiplicative inverse. For example, the multiplicative inverse of the element 2θ is the element $2\theta + 2$.

We can use the multiplication table to check that the multiplicative order of θ in $\mathbb{F}_9$ is 8, which means that θ is a primitive element of $\mathbb{F}_9$. The reader should check that the other root of $p(x)$ is $2\theta + 1$, which turns out to also be a primitive element. Hence the polynomial $p(x)$ is primitive.

With the above examples in mind, we now ask the following question: In general, if p is a prime and $n \geq 2$ is a positive integer, how do we construct a finite field $\mathbb{F}_{p^n}$ with p^n elements?

For example, in our earlier field of order 2^2, note that $p(x) = x^2 + x + 1$, so if θ is a root of $p(x)$, i.e., if $p(\theta) = 0$, then $\theta^2 + \theta + 1 = 0$. Hence we have that $\theta^2 = -\theta - 1 = \theta + 1$. Note that $-1 = 1$, since we are working modulo 2. In this field of four elements, what is θ^3? Note that

$$\theta^3 = \theta(\theta^2) = \theta(\theta + 1) = \theta^2 + \theta = (\theta + 1) + \theta = 1.$$

In the field $\mathbb{F}_{2^2}$ constructed above, what is the element θ^5? We note that

$$\theta^5 = \theta^2(\theta^2)\theta = (\theta + 1)(\theta + 1)\theta = \theta(\theta^2 + 2\theta + 1) = \theta(\theta + 1 + 1) = \theta^2 = \theta + 1.$$

Now consider the field $\mathbb{F}_{3^2}$ containing 3^2 elements discussed earlier. Here the irreducible polynomial used to construct the field was $p(x) = x^2 + x + 2$, so that if θ is a root of $p(x)$, then θ satisfies $\theta^2 + \theta + 2 = 0$, i.e.,

$$\theta^2 = -\theta - 2 = 2\theta + 1,$$

since we are calculating the coefficients modulo 3.

Let p be a prime and let $n \geq 2$ be a positive integer. To construct a finite field $\mathbb{F}_{p^n}$ containing exactly p^n distinct elements, we may proceed as follows. We first consider a monic irreducible polynomial $p(x)$ of degree n over the field $\mathbb{Z}_p$ of integers modulo the prime p. Let θ be a root of the primitive polynomial $p(x)$ of degree n over the field of p elements where p is a prime.

The elements of the field $\mathbb{F}_{p^n}$ can then be viewed as the set of elements

$$\mathbb{F}_{p^n} = \{a_0 + a_1\theta + \cdots + a_{n-1}\theta^{n-1} \mid a_i \in \mathbb{Z}_p \text{ for } i = 0, 1, \ldots, n - 1\}. \quad (5.2)$$

Notice that there are exactly p^n distinct elements in this set since there are p choices for each of the values $a_0, \ldots, a_{n-1}$ and there are n different coefficients.

How do we add these elements? We simply add them as polynomials in θ, reducing the coefficients modulo the prime p. Thus

$$(a_0 + a_1\theta + \cdots + a_{n-1}\theta^{n-1}) + (b_0 + b_1\theta + \cdots + b_{n-1}\theta^{n-1})$$

$$= (a_0 + b_0) + (a_1 + b_1)\theta + \cdots + (a_{n-1} + b_{n-1})\theta^{n-1},$$

where for $i = 0, 1, \ldots, n - 1$, $a_i + b_i$ is computed modulo the prime p.

The reader should note that with this operation of addition, we recover the addition tables for the fields of orders 2^2 and 3^2 given earlier.

Hence, addition of field elements is very easy. In contrast, multiplication of field elements will be seen to be quite detailed and can be tedious, especially in large fields.

To multiply two field elements, we first multiply them as polynomials in θ. We then reduce the coefficients modulo the prime p. We also reduce the powers of θ using the irreducible polynomial $p(x)$ until all exponents are less than the degree of the polynomial $p(x)$.

We note that when performing this multiplication, we will often have powers of θ that are greater than $n - 1$. These powers must be reduced as follows.

Assume that the irreducible polynomial used to construct the finite field $\mathbb{F}_{p^n}$ is given by

$$p(x) = x^n + c_{n-1}x^{n-1} + \cdots + c_1 x + c_0,$$

where each $c_i \in \mathbb{F}_p, i = 0, 1, \ldots, n - 1$.

Let θ be a root of $p(x)$ so that

$$\theta^n + c_{n-1}\theta^{n-1} + \cdots + c_1\theta + c_0 = 0$$

and hence

$$\theta^n = -c_{n-1}\theta^{n-1} - \cdots - c_1\theta - c_0.$$

Now consider the product of two field elements

$$(a_0 + a_1\theta + \cdots + a_{n-1}\theta^{n-1}) \times (b_0 + b_1\theta + \cdots + b_{n-1}\theta^{n-1}).$$

We now replace θ^n by $-c_{n-1}\theta^{n-1} - \cdots - c_1\theta - c_0$. We continue to replace powers of θ like θ^m with $m \geq n$ by the above expression for θ^n until all of the powers of θ in the product are less than n as in (5.2).

In our earlier example of the field with 2^2 elements, we note that $p(x) = x^2 + x + 1$, so if θ is a root of $p(x)$, then $\theta^2 + \theta + 1 = 0$, so that $\theta^2 = -\theta - 1 = \theta + 1$. Note that $-1 = 1$ since we are working modulo 2.

In our previous calculations we have used linear combinations of the elements

$$1, \theta, \theta^2, \ldots, \theta^{n-1}$$

to construct the field elements in a finite field $\mathbb{F}_{p^n}$. Here θ is a root of an irreducible polynomial of degree n over the field $\mathbb{Z}_p$. In fact, as indicated in [13], we can find an irreducible polynomial of degree n over $\mathbb{Z}_p$ that is a primitive polynomial. Hence we can assume that θ is a primitive element.

As the reader can check, the elements $1, \theta, \theta^2, \ldots, \theta^{n-1}$ form a basis for the corresponding vector space; see Chapter 6 for a discussion of vector spaces.

We are really just using the fact that the elements $1, \theta, \theta^2, \ldots, \theta^{n-1}$ form a basis of the finite field $\mathbb{F}_{p^n}$. Hence addition is very easy, as we now point out.

When adding finite field elements $a = a_0 + a_1\theta + \cdots + a_{n-1}\theta^{n-1}$ and $b = b_0 + b_1\theta + \cdots + b_{n-1}\theta^{n-1}$ we see that

$$a + b = (a_0 + b_0) + (a_1 + b_1)\theta + \cdots + (a_{n-1} + b_{n-1})\theta^{n-1},$$

where for $i = 0, 1, \ldots, n - 1$, $a_i + b_i$ is calculated modulo the prime p.

Now we consider multiplication of field elements. Assume that θ is a primitive element in the field so that every non-zero element in the field can be expressed as a power of θ. Suppose for example that $a = \theta^t$ and $b = \theta^r$. Then

$$ab = \theta^t\theta^r = \theta^{t+r}.$$

It is difficult, however, to find the power s of θ such that $\theta^t + \theta^r = \theta^s$.

This illustrates that when doing finite field arithmetic, one of the operations of addition and multiplication is usually quite easy, while the other may involve considerable work.

5.2 Exercises

1. Construct the addition and multiplication tables for the ring

$$\mathbb{Z}_2[x]/(x^2 + x).$$

   Determine whether or not this ring is a field.

   We recall from the discussion earlier in this chapter that by the notation $\mathbb{Z}_2[x]/(x^2 + x)$, we mean all polynomials of degree less than 2 whose coefficients are in the field $\mathbb{Z}_2$. If θ is a root of this polynomial, then $\theta^2 + \theta = 0$ so that over $\mathbb{Z}_2$, $\theta^2 = -\theta = \theta$. The four distinct elements are thus $0, 1, \theta, \theta^2$.

2. Construct a field with 8 elements and determine the multiplicative order of each non-zero element in the field.

3. Construct a field of order 16 and determine the multiplicative order of each non-zero element in the field.

4. If p is a prime then the field $\mathbb{Z}_p$ is the same (isomorphic) as the ring $\mathbb{Z}_p$ of integers modulo p. Explain why the field $\mathbb{F}_4$ is not the same as the ring $\mathbb{Z}_4$ of integers modulo 4. In fact, it turns out that if $m > 1$, the field $\mathbb{F}_{p^m} \neq \mathbb{Z}_{p^m}$, the ring of integers modulo p^m. Explain why.

5. Determine the multiplicative order of each non-zero element in $\mathbb{Z}_{17}$.

5.3   Properties of finite fields

In this section we discuss several important properties of finite fields. We begin with the following:

Lemma 5.6 *If F is a finite field with q elements and $a \in F, a \neq 0$, then $a^{q-1} = 1$, and thus $a^q = a$, for all a in F.*

Proof: The result is certainly true when a is zero. If a is not zero, we know from the definition of a field that a is a unit in the finite field F. There are $q - 1$ units in F, so by Lagrange's Theorem (Theorem 2.13) the multiplicative order of a in F must divide $q - 1$. Therefore $a^{q-1} = 1$ and hence $a^q = a$. ∎

Note that this result provides a generalization of Fermat's Theorem (Theorem 1.22) that if a prime p does not divide a, then $a^{p-1} \equiv 1 \pmod{p}$.

It is easy to see from the previous lemma that the multiplicative inverse of any non-zero element a in a field of order q is a^{q-2}. This follows from the simple calculation $a^{q-2}a = a^{q-1}$ and the definition of an inverse of the non-zero element a.

It turns out (though we will not prove this fact here) that a field of a given order is unique up to field isomorphism (see Definition 3.9 of a ring isomorphism). Thus we may speak of **the** finite field of a particular order q, and we write $\mathbb{F}_q$ to denote this field.

Another common notation for a field of order q is $GF(q)$, where G stands for Galois and F stands for field. This notation is used in honor of Evariste Galois (1811 – 1832), who in 1830 was the first person to seriously study properties of general finite fields (fields with a prime power but not a prime number of elements). We will use the notation $\mathbb{F}_q$ in this book to denote a finite field of order q, i.e., a finite field that contains exactly q distinct elements.

We note that when p is a prime, the field $\mathbb{F}_p$ is the same as (isomorphic as a ring) the ring $\mathbb{Z}_p$ of integers modulo p. In Exercise 5.2.4, the reader was asked to show that when $n > 1$, the finite field $\mathbb{F}_{p^n}$ is **not** the same as the ring $\mathbb{Z}_{p^n}$ of integers modulo p^n. The reader should be sure to understand the difference between these two commutative rings in the non-prime case, when one is a field and the other is not.

Our next result is one of the most important results in the study of finite fields.

Theorem 5.7 *The multiplicative group $\mathbb{F}_q^*$ of all non-zero elements of the finite field $\mathbb{F}_q$ is cyclic.*

Proof: The case where $q = 2$ is trivial so we assume that $q \geq 3$. Let $q - 1 = h > 1$ have the prime factorization $\prod_{i=1}^{t} p_i^{r_i}$. For each i consider the polynomial $f_i(x) = x^{h/p_i} - 1$. This polynomial has degree $h/p_i < h$ and thus has at most h/p_i roots; see Chapter 7 where it is shown that over any field, a polynomial of degree n cannot have more than n roots.

Choose a_i, an element of $\mathbb{F}_q$ that is not a root of f_i, so that $a_i \neq 0$. Let $b_i = a_i^{h/p_i^{r_i}}$.

We claim that the multiplicative order of b_i is $p_i^{r_i}$. Clearly $b_i^{p_i^{r_i}} = a_i^h = 1$. Moreover,

$$b_i^{p_i^{r_i-1}} = a_i^{h/p_i} \neq 1.$$

This implies that the order of b_i is in fact $p_i^{r_i}$, as the order must be some power of p_i; if it were a lower power of p_i than r_i, then $b_i^{p_i^{r_i}}$ would equal 1.

Finally, let $b = b_1 \times \cdots \times b_t$ be the product of the field elements $b_i, i = 1, \ldots, t$. The order of b is $q - 1$, because the orders of each of the elements b_i are relatively prime to each other, and hence the order of their product is the product of the orders; (see Exercise 2.4.4). ∎

An element $\theta \in \mathbb{F}_q$ that multiplicatively generates the group $\mathbb{F}_q^*$ of all non-zero elements of the field $\mathbb{F}_q$ is called a **primitive element**.

Remark 5.8 *Let θ be a primitive element of a finite field F. Then every non-zero element of F can be written as a power of θ. This representation makes* *multiplication of field elements very easy to compute. For example, if $a = \theta^c$ and $b = \theta^d$ for integers c and d, then we have*

$$ab = \theta^c \theta^d = \theta^{c+d}.$$

A subset S of a finite field F is a **subfield** of F if S is itself a field with respect to the same two operations used in the field F; see Definition 4.3 for the definition of a subfield. It turns out that subfields of finite fields are very easy to describe and construct. The following result characterizes the subfield structure in finite fields. While the proof is not too difficult, we will omit the proof and refer the reader to [25] for details. We first state the result, and then provide some illustrations of its use.

Theorem 5.9 *(Subfield structure) Let F be a finite field with p^n elements. Every subfield of F contains p^m elements for some positive integer m dividing n. Conversely, for any positive integer m dividing n, there is a unique subfield of F containing p^m elements.*

As an illustration, the finite field $\mathbb{F}_{2^2}$ contains only two subfields; namely the binary field $\mathbb{Z}_2$ of integers modulo 2 and the field $\mathbb{F}_{2^2}$ itself.

Similarly the field $\mathbb{F}_{3^2}$ containing nine elements has two subfields, namely the field $\mathbb{Z}_3$ of integers modulo 3 and the field $\mathbb{F}_{3^2}$ itself.

By Theorem 5.9, the field $\mathbb{F}_{3^4}$ contains three subfields, namely fields with $3, 3^2 = 9$, and $3^4 = 81$ elements.

In general, the theorem shows that a field with p^n elements contains a number of subfields that is the same as the number of divisors of the positive integer n. Here we are including the two trivial divisors 1 and n of the positive integer n.

How does one find the elements that actually live in a particular subfield? Recall that the multiplicative group of all non-zero elements in a finite field is cyclic (see Theorem 5.7) . Let g be a generator, i.e., let g be a primitive element in the field of order p^m.

Let m be a divisor of n. The reader should check that the elements in the subfield of order p^m are the elements a for which

$$a^{p^m} = a.$$

Recall that if $a \in \mathbb{F}_q$ then from Lemma 5.6 we know that $a^q = q$. The converse also holds, since, as the reader should check, the set of field elements

$$S = \{a \mid a^q = a\}$$

forms a subfield.

Recall from our group theory discussion that every subgroup H of a cyclic group G is cyclic; see Theorem 2.7. In addition, the subgroup H of the group G is generated by the element g^k where k is the smallest power so that $g^k \in H$. Moreover, if a finite group G with n elements is generated by an element g, then the element g^t will also generate the group G if t and n are relatively prime.

We can apply these results to our study of subfields of finite fields. If the field $\mathbb{F}_{p^m}$ is a subfield of the field $\mathbb{F}_{q^n}$ so that by Theorem 5.9 the integer m divides n, then the larger field $\mathbb{F}_{p^n}$ is often called an **extension field** of the field $\mathbb{F}_{p^m}$. In fact, the extension field may be viewed as a vector space over the subfield. Recall from Chapter 6 on vector spaces that the dimension of the extension field $\mathbb{F}_{p^n}$ over the subfield $\mathbb{F}_{p^m}$ is the integer n/m.

For example, the field $\mathbb{F}_{2^2}$ is a vector space of dimension 2 over the subfield $\mathbb{Z}_2$. Recall in our construction of the field $\mathbb{F}_{2^2}$, we noted that elements of the field with four elements had the form $a + b\theta$ with $a, b \in \mathbb{Z}_2$. In terms of vector spaces, this is simply saying that the elements 1 and θ form a basis of the field $\mathbb{F}_{2^2}$ over the subfield $\mathbb{Z}_2$. Here θ satisfies $\theta^2 + \theta + 1 = 0$.

In the field $\mathbb{F}_{2^2}$, θ is a primitive element so that the elements 0 and $\theta^3 = 1$ form the subfield $\mathbb{Z}_2$, The elements θ and θ^2 are in the field $\mathbb{F}_{2^2}$ but they are not in the subfield $\mathbb{Z}_2$ since $\theta^2 = \theta + 1$ and $(\theta^2)^2 = \theta^4 = \theta$, since we have that $\theta^3 = 1$.

Similarly the field $\mathbb{F}_{3^2}$ is a vector space of dimension 2 over the subfield $\mathbb{Z}_3$. Here the elements 1 and θ also form a basis where θ now satisfies $\theta^2 + \theta + 2 = 0$ so that $\theta^2 = -\theta - 2 = 2\theta + 1$. The element θ is a primitive element so that the subfield $\mathbb{Z}_3$ consists of the elements

$$0, \theta^4 = -1 = 2, \theta^8 = 1.$$

Note that

$$(\theta^4)^3 = \theta^{12} = \theta^4.$$

so that $\theta^4 \in \mathbb{Z}_3$. Similarly, $(\theta^8)^3 = \theta^8$ so that $\theta^8 = 1 \in \mathbb{Z}_3$.

The trace function is of fundamental importance in the study of finite field theory. It is also useful in various applications of finite fields. We now define the trace function as follows:

Let $K = \mathbb{F}_q$ and $F = \mathbb{F}_{q^m}$. For $\alpha \in F$, we define the **trace** of α over K as

$$\mathrm{Tr}_{F/K}(\alpha) = \alpha + \alpha^q + \cdots + \alpha^{q^{m-1}}.$$

As an illustration, let $K = \mathbb{Z}_2$ and $F = \mathbb{F}_{2^4}$. Then

$$\mathrm{Tr}_{F/K}(\alpha) = \alpha + \alpha^2 + \alpha^4 + \alpha^8.$$

For $K = \mathbb{F}_4$ and $F = \mathbb{F}_{16}$ we have $\mathrm{Tr}_{F/K}(\beta) = \beta + \beta^4$.

We now ask the reader to prove some of its properties. These include

1. $\mathrm{Tr}_{F/K}(\alpha + \beta) = \mathrm{Tr}_{F/K}(\alpha) + \mathrm{Tr}_{F/K}(\beta)$ for $\alpha, \beta \in F$;

2. $\mathrm{Tr}_{F/K}(c\alpha) = c\mathrm{Tr}_{F/K}(\alpha)$ for $\alpha \in F$;

3. The trace function is a linear map from F onto K;

4. $\mathrm{Tr}_{F/K}(\alpha) = m\alpha$ for $\alpha \in K$;

5. $\mathrm{Tr}_{F/K}(\alpha^q) = \mathrm{Tr}_{F/K}(\alpha)$ for $\alpha \in F$.

5.3 Exercises

1. For the primes $p = 2, 3$, and 5, and for the positive integers $n = 2, 3$, and 4, determine the number $N_p(n)$ of monic irreducible polynomials of degree n over the field $\mathbb{Z}_p$.

2. Show that the sum of all elements of a finite field is 0, except for the field $\mathbb{Z}_2$.

3. Construct the field $\mathbb{F}_{2^4}$ containing 16 elements and determine the elements in each of the subfields of the field $\mathbb{F}_{2^4}$. Use the polynomial $x^4 + x + 1$ to construct the field $\mathbb{F}_{2^4}$.

4. Construct the field $\mathbb{F}_{3^3}$ containing 27 elements and determine the elements in each of the subfields of the field $\mathbb{F}_{3^3}$. Use the polynomial $x^3 + 2x + 1$ to construct the field $\mathbb{F}_{3^3}$.

5. Show that every finite field of non-prime order must contain at least two subfields.

6. Determine all finite fields which have exactly two subfields.

7. Determine the number of subfields of the field $\mathbb{F}_{2^{10}}$.

5.4   Polynomials over finite fields

In this section we discuss various properties related to polynomials over finite fields. Our first result is that every function defined from a finite field to itself can be represented by a polynomial with coefficients in that finite field.

This property is an extremely important property of finite fields. In fact, it characterizes finite fields in the sense that finite fields are the only finite commutative rings with a unity element with the property that every function defined from the ring to itself can be realized by a polynomial with coefficients in that ring. The next result tells us how to obtain a polynomial representing a given function over a finite field.

Theorem 5.10 *(Lagrange Interpolation Formula) Every function $f : \mathbb{F}_q \to \mathbb{F}_q$ can be represented by a unique polynomial over $\mathbb{F}_q$ of degree at most $q - 1$.*

Proof: By representing a function f by a polynomial P_f, we mean that $P_f(a) = f(a)$ for each $a \in \mathbb{F}_q$.

Let f be a function from $\mathbb{F}_q$ to itself so that $f : \mathbb{F}_q \to \mathbb{F}_q$. Define a polynomial $P_f(x)$ over $\mathbb{F}_q$, i.e., with its coefficients in the field $\mathbb{F}_q$, by

$$P_f(x) = \sum_{a \in \mathbb{F}_q} f(a)\left[1 - (x - a)^{q-1}\right].$$

Note that the degree of $P_f(x)$ is at most $q - 1$ by its very construction. Note also that $(a - b)^{q-1}$ is equal to 1 if $a \neq b$ and is equal to 0 if $a = b$. A straightforward calculation shows that $P_f(a) = f(a)$ for all $a \in \mathbb{F}_q$, so the polynomial $P_f(x)$ indeed represents the function $f(x)$. ∎

For example, consider the function $f : \mathbb{Z}_3 \to \mathbb{Z}_3$ defined by

$$f(0) = 0, f(1) = 2, f(2) = 1.$$

Using the Lagrange Interpolation Formula from Theorem 5.10, we can calculate

$$P_f(x) = 2[1 - (x - 1)^2] + 1[1 - (x - 2)^2]$$

$$= 2(-x^2 + 2x) + 1(-x^2 + 4x - 3) = -3x^2 + 8x - 3 = 2x,$$

since over the field $\mathbb{Z}_3$, we compute modulo 3.

As a slightly larger example, consider the function defined on the field $\mathbb{Z}_5$ by

$$f(0) = 0, f(1) = f(4) = 1, f(2) = f(3) = 4.$$

By the Lagrange Interpolation Formula we have

$$P_f(x) = [1 - (x - 1)^4] + 4[1 - (x - 2)^4] + 4[1 - (x - 3)^4] + 1[1 - (x - 4)^4],$$

where all calculations are done modulo 5.

We note that

$$(x - a)^4 = x^4 - 4ax^3 + ba^2x^2 - 4a^3x + a^4.$$

Hence we find, after some simplification, that

$$P_f(x) = -x^4 + (-x^4 + 4x^3 - x^2 + 4x) + (-4x^4 + 2x^3 - x^2 + 3x)$$
$$+ (-4x^4 + 3x^3 - x^2 + 2x) + (-x^4 + x^3 - x^2 + x).$$

Upon summing these expressions we obtain

$$P_f(x) = -5x^4 + 10x^3 - 4x^2 + 10x = -4x^2 = x^2.$$

5.4 Exercises

1. Evaluate $f(3)$ for $f(x) = x^{214} + 3x^{152} + 2x^{47} + 2 \in \mathbb{Z}_5[x]$.

2. Is the element 3 a root of the polynomial $x^{12} + x^8 + x^2 + 2$ over the field $\mathbb{Z}_5$ of integers modulo 5?

3. Find all roots of the polynomial $f(x) = x^4 - 2x^2 - 3$ over the field $\mathbb{Z}_5$.

4. How many roots does the polynomial $f(x) = x^3 + x^2 + x + 1$ have in the fields $\mathbb{Z}_2, \mathbb{Z}_3, \mathbb{Z}_5, \mathbb{Z}_7$?

5. Determine the number of functions mapping $\mathbb{F}_q$ to itself, i.e., determine the number of functions $f : \mathbb{F}_q \to \mathbb{F}_q$ for any q.

6. Make lists of all of the polynomials of degrees at most one and of degree at most two over the finite fields $\mathbb{Z}_2$ and $\mathbb{Z}_3$.

7. Use the Lagrange Interpolation Formula to find the polynomial $P_f(x)$ that represents the function $f(x)$ over the field $\mathbb{F}_{2^2}$ of four elements defined by the following:

$$
\begin{aligned}
f(0) &= 0 \\
f(1) &= 1 \\
f(\theta) &= \theta + 1 \\
f(\theta + 1) &= \theta,
\end{aligned}
$$

where θ is a root of the irreducible polynomial $p(x) = x^2 + x + 1$.

8. Use the Lagrange Interpolation Formula to find the polynomial $P_f(x)$ which represents the function $f(x)$ over the field $\mathbb{Z}_5$ of integers modulo 5 which has the property that

$$
\begin{aligned}
f(0) &= 0 \\
f(1) &= 1 \\
f(2) &= 3 \\
f(3) &= 2 \\
f(4) &= 4.
\end{aligned}
$$

9. Prove that $(f(x))^q = f(x^q)$ for any polynomial $f(x) \in \mathbb{F}_q[x]$.

   The property described in this exercise is of great use in finite field calculations.

10. Let $L(x) = \sum_{i=0}^{n} \alpha_i x^{q^i}$, where $\alpha_i \in \mathbb{F}_{q^m}$. A polynomial of this form is called a **linearized polynomial** (another name is q-**polynomial** because the exponents are all powers of q). These polynomials form an important class of polynomials over finite fields because they induce linear functions from $\mathbb{F}_{q^m}$ to $\mathbb{F}_q$. Show that for all $\alpha, \beta \in \mathbb{F}_{q^m}$ and all $c \in \mathbb{F}_q$:

(a) $L(\alpha + \beta) = L(\alpha) + L(\beta)$,

(b) $L(c\alpha) = cL(\alpha)$.

5.5 Permutation polynomials

In this section we briefly discuss permutation polynomials, a class of polynomials that have various applications in combinatorics and cryptography.

A polynomial $f : \mathbb{F}_q \to \mathbb{F}_q$ is a **permutation polynomial** if $f(x)$ induces a bijective mapping on the field $\mathbb{F}_q$. Recall that a bijective mapping is a mapping which is both injective and surjective, i.e., the mapping is both 1-1 and onto.

In Exercise 5.5.2, we ask the reader to show that if $f(x)$ is a permutation polynomial defined on the finite field $\mathbb{F}_q$, then the polynomial $af(x + b) + c$ is also a permutation polynomial for all $a, b, c \in \mathbb{F}_q$, with $a \neq 0$.

Thus, given one permutation polynomial, we can easily generate $q^2(q - 1)$ others since there are $q - 1$ non-zero choices for the element a and q choices for each of the values b and c in the finite field $\mathbb{F}_q$.

Given a polynomial $f(x)$ over $\mathbb{F}_q$, how does one determine if $f(x)$ is actually a permutation polynomial on $\mathbb{F}_q$? One could of course substitute the q field elements into the polynomial and then check if the q image values $f(a)$ are distinct. This, however, is not efficient if q is very large. In fact no efficient test, in terms of the coefficients of the polynomial, is known. We state the following result but refer to Theorem 7.4 of [22] for a proof.

Theorem 5.11 *(Hermite/Dickson criterion)* Let $q = p^m$ where p is a prime. Then a polynomial $f(x)$ over $\mathbb{F}_q$ is a permutation polynomial on $\mathbb{F}_q$ if and only if the following two conditions hold:

1. $f(x)$ has exactly one root in $\mathbb{F}_q$;

2. For each integer t with $1 \leq t \leq q - 2$ and $t \not\equiv 0 \pmod{p}$, the reduced polynomial $(f(x))^t \pmod{x^q - x}$ has degree at most $q - 2$.

We obtain the following corollary by applying the Hermite–Dickson criterion (with $t = (q - 1)/d$) to a polynomial of degree $d > 1$ over $\mathbb{F}_q$.

Corollary 5.12 *There is no permutation polynomial of degree $d > 1$ over $\mathbb{F}_q$ if d divides $q - 1$.*

Theorem 5.13 *The polynomial x^n induces a permutation of $\mathbb{F}_q$ if and only if $\gcd(n, q - 1) = 1$.*

Proof: The proof follows from the fact that the multiplicative group $\mathbb{F}_q^*$ of non-zero elements of $\mathbb{F}_q$ is cyclic and the fact that $\gcd(n, q - 1) = 1$. ∎

We briefly consider another important class of permutation polynomials defined over finite fields.

As will be further discussed in Section 7.3 when dealing with polynomials, if $a \in \mathbb{F}_q$ and $n \geq 2$ is an integer, we define the **Dickson polynomial of degree n and parameter a** by

$$D_n(x, a) = \sum_{i=0}^{\lfloor n/2 \rfloor} \frac{n}{n - i} \binom{n - i}{i} (-a)^i x^{n-2i}.$$

For $n = 0$ we define the Dickson polynomial $D_0(x, a) = 2$ and similarly we define $D_1(x, a) = x$.

We note that Dickson polynomials may be viewed as generalizations of the power or cyclic polynomials x^n since $D_n(x, 0) = x^n$.

The following theorem provides an effective way to determine when the Dickson polynomial $D_n(x, a)$ is a permutation polynomial over the field $\mathbb{F}_q$.

Theorem 5.14 *For any non-zero $a \in \mathbb{F}_q$, the Dickson polynomial $D_n(x, a)$ is a permutation polynomial on $\mathbb{F}_q$ if and only if $(n, q^2 - 1) = 1$.*

We omit the proof of this result and instead refer the reader to Theorem 1.6.21 of [25] or Theorem 7.16 of [22]. We note that when $a \in \mathbb{F}_q$ is non-zero, the Dickson polynomial $D_n(x, a)$ is either a permutation for all a, or it is not a permutation for any a.

5.5 Exercises

1. (i) Construct all permutation polynomials on the field $\mathbb{Z}_2$ of degree at most one.

   (ii) Construct all permutation polynomials on the field $\mathbb{Z}_3$ of degree at most two.

   (iii) Construct all permutation polynomials on the field $\mathbb{Z}_5$ of degree at most four.

2. If $f(x)$ is a permutation polynomial over the field $\mathbb{F}_q$, show that for all $a, b, c \in \mathbb{F}_q$ with $a \neq 0$, the polynomial

$$af(x + b) + c$$

   is also a permutation polynomial over $\mathbb{F}_q$.

3. For an odd prime p, show that the polynomial

$$x^{p-2} + x^{p-3} + \cdots + x^2 + 2x + 1$$

   represents the permutation that consists of the transposition (01) which interchanges 0 and 1, and fixes all other elements of the field $\mathbb{Z}_p$.

4. For non-zero values of a, determine which of the Dickson polynomials $D_n(x, a)$ over the fields $\mathbb{Z}_5$ and $\mathbb{Z}_7$ are permutation polynomials for $1 \le n \le 10$.

5.6 Applications

Finite fields have many applications. In Chapter 8, we will discuss applications of finite fields to error-correcting (algebraic) coding theory. We will now discuss two additional applications. One is a combinatorial application to the construction of sets of mutually orthogonal latin squares. The other is a cryptographic application dealing with the secure distribution of a common key between two parties.

5.6.1 Orthogonal Latin squares

In this subsection we show how to use finite fields and polynomials over finite fields to construct sets of mutually orthogonal latin squares. This is a combinatorial application of finite fields, with a long history of study. In particular, the Swiss–born mathematician Leonhard Euler (1707 – 1783), arguably the most prolific mathematician of the eighteenth century, studied such squares in 1782. See [18] for a historical account of this work.

An $n \times n$ square based upon n distinct symbols is a **latin square of order** n if each of the n symbols occurs exactly once in each row and in each column of the square. Thus

$$\begin{array}{ccc} 0 & 1 & 2 \\ 1 & 2 & 0 \\ 2 & 0 & 1 \end{array}$$

is a latin square of order 3.

This leads to an obvious question: Given a positive integer n, is there a latin square of order n?

The answer to this question is "Yes," because we can take $0, 1, \ldots, n-1$ to be the first row of the square and then to form the second row, shift each element one position to the left, and move the first element so that it is the last element in the second row. Thus the second row of our square becomes $1, 2, \ldots, n-1, 0$. Similarly we continue to shift each row in this manner to obtain the next row, continuing until we have n rows to complete the square. This results in a latin square of order n. When $n = 3$, this process gives the above latin square of order 3.

In fact, for any positive integer n the addition table of the integers modulo n forms a latin square of order n. Hence, there is a latin square of order n for any positive integer n.

Two latin squares of order n are **orthogonal**, if when placing one square on top of the other, we obtain each of the n^2 possible ordered pairs exactly once. In addition, a set of $t \geq 2$ latin squares is **mutually orthogonal** if each pair of distinct squares in the set is orthogonal. In this case we say that we have a set of mutually orthogonal latin squares (MOLS).

In Exercise 5.6.10 the reader will be asked to show that the maximum number of mutually orthogonal latin squares of order n is $n - 1$. Because of this result, a set of t MOLS of order n is said to be **complete** if $t = n - 1$.

For example, the pair of squares

$$
\begin{array}{ccc}
0 & 1 & 2 \\
1 & 2 & 0 \\
2 & 0 & 1
\end{array}
\qquad
\begin{array}{ccc}
0 & 1 & 2 \\
2 & 0 & 1 \\
1 & 2 & 0
\end{array}
$$

is a complete set of two MOLS of order 3.

We will soon see that these latin squares can be very simply constructed by using linear polynomials over the finite field $\mathbb{Z}_3$ of integers modulo 3.

The following latin squares of order 4 are obtained from the finite field $\mathbb{F}_4 = \{0, 1, \theta, \theta + 1\}$ where θ is a root of the irreducible polynomial $p(x) = x^2 + x + 1$ over the field $\mathbb{Z}_2$.

$$
\begin{array}{cccc}
0 & 1 & \theta & \theta + 1 \\
1 & 0 & \theta + 1 & \theta \\
\theta & \theta + 1 & 0 & 1 \\
\theta + 1 & \theta & 1 & 0
\end{array}
$$

$$
\begin{array}{cccc}
0 & 1 & \theta & \theta + 1 \\
\theta & \theta + 1 & 0 & 1 \\
\theta + 1 & \theta & 1 & 0 \\
1 & 0 & \theta + 1 & \theta
\end{array}
$$

$$
\begin{array}{cccc}
0 & 1 & \theta & \theta + 1 \\
\theta + 1 & \theta & 1 & 0 \\
1 & 0 & \theta + 1 & \theta \\
\theta & \theta + 1 & 0 & 1
\end{array}
$$

is a complete set of three MOLS of order 4.

If the reader prefers to view the set of three MOLS of order 4 as consisting of the numbers $0, 1, 2, 3$, one could, in each of the three squares, replace $0, 1, \theta, \theta + 1$ by the numbers $0, 1, 2, 3$. Thus the values 0 and 1 would stay the same, but each θ would be replaced by a 2 and each $\theta + 1$ would be replaced by a 3.

Making these substitutions would yield the squares

$$
\begin{array}{cccc}
0 & 1 & 2 & 3 \\
1 & 0 & 3 & 2 \\
2 & 3 & 0 & 1 \\
3 & 2 & 1 & 0
\end{array}
\qquad
\begin{array}{cccc}
0 & 1 & 2 & 3 \\
2 & 3 & 0 & 1 \\
3 & 2 & 1 & 0 \\
1 & 0 & 3 & 2
\end{array}
\qquad
\begin{array}{cccc}
0 & 1 & 2 & 3 \\
3 & 2 & 1 & 0 \\
1 & 0 & 3 & 2 \\
2 & 3 & 0 & 1
\end{array}
.
$$

Recall from our earlier discussion in this chapter that for any prime power, say q, there is a finite field containing exactly q distinct elements. With these facts in mind, we are led to a significant result on MOLS proven by the Indian mathematician R. C. Bose (1901 – 1987), who made significant contributions in mathematics and statistics throughout his career. In particular, in [3] Bose proved the following result.

Theorem 5.15 *(Bose) Let q be a prime power and let $a \in \mathbb{F}_q, a \neq 0$, where $\mathbb{F}_q$ is the finite field of order q. Then the polynomials $ax + y$ give a complete set of $q - 1$ mutually orthogonal latin squares of order q.*

Proof: We first label (in any order) the rows and columns of a $q \times q$ latin square with the elements of the finite field $\mathbb{F}_q$. The rows and columns need not be labelled in the same order although using the same ordering for the rows and columns can make the arithmetic and structure easier to follow.

From the polynomial $ax + y$ we form a $q \times q$ square by placing the field element $ax + y$ at the intersection of row x and column y of the a-th square.

We first show that such a square is indeed a latin square of order q. We will show that each row and each column contains q distinct elements.

In row x, assume that

$$ax + y_1 = ax + y_2.$$

Then by subtracting ax from both sides we see that $y_1 = y_2$ and so row x must contain distinct elements.

Similarly by column, assume that in column y we have

$$ax_1 + y = ax_2 + y.$$

Then we must have $ax_1 = ax_2$. Since a is a non-zero field element, we can divide by a to show that $x_1 = x_2$ and hence the square must have distinct elements in this column. Thus the square obtained from the polynomial $ax + y$ is a latin square of order q.

Now consider two latin squares constructed from two linear polynomials $ax + y$ and $bx + y$ where a and b are distinct non-zero elements of the finite field $\mathbb{F}_q$.

If the two squares are not orthogonal, there must then be two cells in the square obtained from $ax + y$ that have the same element, say c, and in the same cells of the square obtained from the polynomial $bx + y$ we must have a common value, say d. Thus we would have the pair (c, d) occurring twice when the two squares are placed on top of each other.

Then we would have the pair of simultaneous equations

$$ax_1 + y_1 = ax_2 + y_2$$

$$bx_1 + y_1 = bx_2 + y_2.$$

Thus upon subtraction we would have

$$(a - b)x_1 = (a - b)x_2.$$

Since $a \neq b$, we can divide by the field element $a - b$ to show that $x_1 = x_2$. But then we must also have $y_1 = y_2$. Hence the pair of squares is indeed orthogonal and the proof is complete. ∎

The reader should observe that the pair of MOLS of order 3 given earlier arises from the polynomials $x + y$ and $2x + y$ over the field $\mathbb{Z}_3$ of integers modulo 3. In the exercises, the reader will be asked to construct various sets of mutually orthogonal latin squares.

There is a famous unsolved conjecture concerning when there is a complete set of $n - 1$ MOLS of order n. It is known as the **Prime Power Conjecture** and can be stated as follows:

Conjecture 5.16 *(Prime Power Conjecture) There is a complete set of $n - 1$ MOLS of order n if and only if n is a prime power.*

For a short discussion of this conjecture where it has been given the name "The Next Fermat Problem," see [24].

We note that it is known that there is no pair of MOLS of order 2 or 6. The reader should verify this fact for squares of order 2.

For squares of order 6, Euler conjectured that there is no pair of MOLS of order 6. This remained an unsolved problem until 1900 when Tarry (see pages 140 + of [6]) proved, by a very exhaustive search, and without the aid of a computer, that there cannot be a pair of MOLS of order 6.

The reader should see Stinson [31] for a short proof of this fact, which does not require the use of a computer. We note that even with a computer, the proof that there is no pair of MOLS of order 6 is non-trivial as there are $6!5!9408$ or $812,851,200$ latin squares of order 6 [6].

After 6, the next non-prime power is 10. What happens for latin squares of order 10? We really don't know much about this case except that there cannot be a complete set of 9 MOLS of order 10; see [18].

It is known that there are pairs of orthogonal latin squares of order 10; see [18]. But it is not known whether there is a set of 3 MOLS of order 10, see [18]. Even with the advent of very fast modern computers, we have not been able to resolve the question of whether there is a set of 3 MOLS of order 10.

5.6.2   Diffie/Hellman key exchange

Suppose that two parties want to be able to share a common cryptographic key without eavesdroppers being able to determine the key. How could one do that and where might such a system be used? One place that such a system was used was during the Cold War when the United States and the Soviet Union wanted to establish their hotline.

Let's call the two users A and B. In cryptographic circles these letters often stand for Alice and Bob. First, assume that they agree to choose a large prime p to serve as the basis for the construction of their key. How large should p be? For very important issues that involve serious security issues, p would be very large, at least 100 base 10 digits or likely more in length.

Then A and B agree to share a primitive root g in the field $\mathbb{Z}_p$. Recall the difficulty of first obtaining such a large prime, and then the difficulty of locating a primitive root for that prime; see Remark 2.8. Nevertheless, these obstacles can be overcome, and so let us assume that both A and B now share the prime p and the primitive element g in the field of integers modulo the prime p.

Next, A and B both choose secret keys, say a and b. These keys are integers in the range $2 \leq a, b \leq p - 2$. For security reasons, it is essential that A keeps her key a to herself and that B also keeps his key b to himself.

Next A obtains the least non-negative residue c of the value g^a calculated modulo p. She then sends the value c to Bob who calculates c^b modulo p. Notice the net effect is that Bob (without knowing a) then has the value

$$c^b \equiv (g^a)^b \equiv g^{ab} \pmod{p}.$$

Similarly, B calculates the value g^b modulo p, and obtains the least non-negative residue, say d. This value d is then sent to A. Upon receiving the value d, A then calculates

$$d^a \equiv (g^b)^a \equiv g^{ba} \pmod{p}.$$

Notice that the two values

$$g^{ab} \equiv g^{ba} \pmod{p}$$

are the same modulo the prime p. Thus A and B have now both calculated this common value, known as the common key.

This system was first discovered by W. Diffie and M. E. Hellman in [8].

Example 5.17 *We now illustrate the Diffie/Hellman key-exchange system with a small example. Of course there is no security in our example, as the numbers are far too small. Nevertheless, the example will hopefully help to make you realize that the system really does work.*

Let $p = 11$ and choose the primitive element 2. You should convince yourself that 2 really is a primitive element modulo 11.

Assume that A chooses her secret key to be $a = 8$ and B chooses his secret key to be $b = 9$. To share a common key, A and B calculate as follows.

The user A calculates

$$2^8 \equiv 2^4 2^4 \equiv 5(5) \equiv 3 \pmod{11}.$$

Upon receiving the value 3 from A, Bob calculates

$$3^9 \equiv 3^4 3^4 3^1 \equiv 4(4)(3) \equiv 4 \pmod{11}.$$

Similarly, B calculates

$$2^9 \equiv 2^4 2^4(2) \equiv 5(5)(2) \equiv 6 \pmod{11}$$

and sends the value 6 to A.

The user A then calculates

$$6^8 \equiv (6^2)^4 \equiv 3^4 \equiv 4 \pmod{11}.$$

Thus A and B now both have the common key 4.

In actual practice, this common key is often used to set up another cryptosystem to securely transmit information.

One could also employ the Diffie/Hellman Key Exchange Scheme over a general finite field of order $q = p^n$ where p is a prime and $n > 1$ is a positive integer. The system is set up the same way starting with a primitive element g in $\mathbb{F}_q$. Party A now chooses her secret key a with $2 \leq a \leq q - 2$ and party B chooses his secret key b with $2 \leq b \leq q - 2$.

One drawback of this general finite field method where we work over the extension field is that all calculations must then be done as extension field calculations. Such calculations are more costly than doing the calculations in $\mathbb{Z}_p$ modulo p where p is a large prime.

The reader may wonder how secure the Diffie/Hellman scheme really is. We note that the elements $c = g^a \in \mathbb{Z}_p$ and $d = g^b \in \mathbb{Z}_p$ are sent across the communication line. The public has access to the values c and d (but not a and b). An unauthorized attacker also knows the prime p and the primitive element g.

So the real problem is the following: given c and g, can the attacker obtain the secret key a where $c = g^a$? This question is known as the **Discrete Logarithm Problem**, and is thought to be very difficult to solve for very large primes p.

We now state this problem for prime fields $\mathbb{Z}_p$.

Definition 5.18 *(Discrete Logarithm Problem) Let g be a primitive element in $\mathbb{Z}_p$ where p is a prime. The* **Discrete Logarithm Function** *is the unique function $\log_g : \mathbb{Z}_p^* \to \{0, 1, \ldots, p - 2\}$, which makes the equation*

$$a = g^{\log_g(a)}$$

hold for every $a \in \mathbb{Z}_p^*$.

The function $\log_g$ is thought to be very difficult to compute, although its inverse $a \to g^a$ is very easy to compute. If an efficient method for computing discrete logarithms were discovered, the Diffie/Hellman Key Exchange would no longer be secure. It is known that computing discrete logarithms in $\mathbb{F}_q$ is of about the same level of computational difficulty as factoring an RSA modulus $n = rs$ (r and s primes) when q and n are of comparable size.

The following formula for the Discrete Logarithm Problem is given in [27].

Theorem 5.19 *Let p be a prime and let g be a primitive element modulo p.*
Then

$$log_g(a) = -1 + \sum_{j=1}^{p-2} \frac{a^j}{g^{-j} - 1}.$$

The above formula may appear to be a bit confusing. We have indicated that the Discrete Logarithm Problem is believed to be difficult to solve for a large prime p. And yet the above expression provides an exact formula for the solution! It turns out that computationally, it is not feasible to calculate the sum in the expression.

The reason this formula is not useful computationally is that it involves $p - 2$ terms. If p is a really large prime, even with today's fast computers, one cannot calculate such a large number of terms.

Before closing this chapter on finite fields, we mention two other books that discuss various aspects of finite fields; see [21] and [26].

5.6 Exercises

1. Construct a complete set of 3 MOLS of order 4.

2. Construct a complete set of 4 MOLS of order 5.

3. Construct 2 MOLS of order 8.

4. Construct 2 MOLS of order 9.

5. Construct 2 MOLS of order 12. Hint: Try to glue together pairs of MOLS of orders 3 and 4.

6. Assume that $p = 13$. First, find a primitive element modulo 13. Now assume that A and B are using the Diffie/Hellman key exchange system to interchange a common key.

   If A chooses her secret key to be $a = 3$ and B chooses his secret key to be $b = 5$, determine the common key.

7. Repeat Exercise 5.6.6 with $p = 17$ with the same values for the secret keys a and b.

8. Repeat Exercise 5.6.6 with $p = 23$ and $a = 2$ and $b = 3$.

9. In a Diffie/Hellman system modulo a prime p, how many choices are there for the secret key a for the user A?

10. Show that for any integer $n \geq 2$, there cannot be more than $n - 1$ MOLS of order n.

Chapter 6

Vector Spaces

6.1   Definition and examples

In this chapter, we describe some basic properties of vector spaces. These algebraic structures occur in many different areas of mathematics. For example, they are very useful structures in algebraic coding theory for the error-free transmission of information, as will be illustrated in Chapter 8.

Given a field F, a **vector space** V over F is an additive Abelian group for which we also have a "scalar multiplication" that must satisfy the following properties.

- V1: for each $\mathbf{v} \in V$ and $\lambda \in F$, $\lambda\mathbf{v} \in V$;

- V2: for each $\mathbf{v} \in V$ and $\lambda, \mu \in F$, $(\lambda\mu)\mathbf{v} = \lambda(\mu\mathbf{v}) \in V$;

- V3: for each $\mathbf{v} \in V$, $1\mathbf{v} = \mathbf{v}$;

- V4: for each $\mathbf{v} \in V$ and $\lambda, \mu \in F$, $(\lambda + \mu)\mathbf{v} = \lambda\mathbf{v} + \mu\mathbf{v}$;

- V5: for each $\mathbf{u}, \mathbf{v} \in V$ and $\lambda \in F$, $\lambda(\mathbf{u} + \mathbf{v}) = \lambda\mathbf{u} + \lambda\mathbf{v}$.

We note that the "1" in property V3 refers to the multiplicative identity in the field F.

The elements of V are called **vectors** while the elements of F are called **scalars**. Throughout this chapter, we denote vectors in bold type to distinguish them from scalars.

There are numerous examples of structures that form vector spaces. We now provide a few examples of such structures.

Example 6.1 *Let $V = \mathbb{R}^2$ be the usual xy-plane. We define addition of vectors via*

$$(x_1, x_2) + (y_1, y_2) = (x_1 + y_1, x_2 + y_2)$$

and scalar multiplication via

$$\lambda(x_1, x_2) = (\lambda x_1, \lambda x_2).$$

The set V forms a vector space over the field $\mathbb{R}$ of real numbers.

Example 6.2 *One could similarly form vector spaces by considering $\mathbb{R}^3$ or even $\mathbb{R}^n$ over the field of real numbers for any positive integer n.*

In this more general setting, we define

$$(x_1, \ldots, x_n) + (y_1, \ldots, y_n) = (x_1 + y_1, \ldots, x_n + y_n)$$

and

$$\lambda(x_1, \ldots, x_n) = (\lambda x_1, \ldots, \lambda x_n).$$

The next example involves the field of complex numbers and the field of real numbers, and provides an example where a field forms a vector space over a subfield.

Example 6.3 *The complex numbers form a vector space over the field of real numbers. This follows from the complex number calculations*

$$(a + bi) + (c + di) = (a + c) + (b + d)i$$

and for any real number λ

$$\lambda(a + bi) = \lambda a + \lambda bi.$$

Remark 6.4 *We note, however, that the real numbers do not form a vector space over the field of complex numbers. See Exercise 6.1.4.*

As the next three examples indicate, matrices can also be used to form vector spaces over any field.

Example 6.5 *The set of all $n \times n$ matrices over the real numbers, i.e., the set of all $n \times n$ matrices with real numbers as entries, forms a vector space over the field of real numbers where addition of matrices is the usual matrix addition defined as follows.*

If $A = (a_{ij})$ and $B = (b_{ij})$ are two $n \times n$ matrices with real number entries, then the sum $A + B$ is defined by

$$A + B = (a_{ij}) + (b_{ij}) = (a_{ij} + b_{ij}), 1 \leq i, j \leq n.$$

Scalar multiplication of matrices is defined by

$$\lambda A = \lambda(a_{ij}) = (\lambda a_{ij}).$$

Example 6.6 *In an analogous way we could form a vector space over the real numbers by considering the set of all $m \times n$ matrices, i.e, matrices with m rows and n columns whose entries are real numbers. Addition and scalar multiplication are then defined in the same way, namely by defining the addition and scalar multiplication elementwise as in Example 6.4.*

Example 6.7 *Given any field F, finite or infinite, the sets of $n \times n$ or $m \times n$ matrices with entries from the field F form vector spaces with definitions of vector addition and scalar multiplication defined as above.*

As the following examples indicate, polynomials can also be used to form various vector spaces over any field.

Example 6.8 *The set $\mathbb{R}[x]$ of all polynomials whose coefficients are real numbers with addition defined by adding polynomials in the usual way is a vector space over the field of real numbers. Scalar multiplication by a real number λ is defined by simply multiplying each coefficient of a polynomial by the scalar λ.*

One can also consider the set of all polynomials with real coefficients whose degree is at most n. Using the usual operations of polynomial addition and scalar multiplication, this set of polynomials also forms a vector space over the real numbers. Similarly, one could form further vector spaces by considering the same sets of polynomials over any field F.

6.1 Exercises

1. Do the real numbers form a vector space over the rational numbers?

2. Do the complex numbers form a vector space over the rational numbers?

3. Do the rational numbers form a vector space over either the real numbers or the complex numbers?

4. Explain why the real numbers do not form a vector space over the complex numbers.

5. Let S be the set of all polynomials of degree 2 with rational coefficients. Does S form a vector space over the field $\mathbb{Q}$ of rational numbers?

6. If $a, b \in F$ a field, $\mathbf{x} \in V, \mathbf{x} \neq \mathbf{0}$ a vector space, and if $a\mathbf{x} = b\mathbf{x}$, then show that $a = b$.

7. If $a \in F, a \neq 0$ a field, and $\mathbf{x}, \mathbf{y} \in V$ a vector space with $a\mathbf{x} = a\mathbf{y}$, show that $\mathbf{x} = \mathbf{y}$.

8. Consider the set V of all 2×2 matrices over the field of real numbers with the usual addition of matrices. Define scalar multiplication by

$$\lambda \begin{pmatrix} a & b \\ c & d \end{pmatrix} = \begin{pmatrix} \lambda a & b \\ c & \lambda d \end{pmatrix}.$$

   Does V form a vector space over the field of real numbers?

9. Let V and W be vector spaces over a field F. A mapping

$$f : V \to W$$

   is a **linear mapping** if

$$f(\mathbf{x} + \mathbf{y}) = f(\mathbf{x}) + f(\mathbf{y})$$

   for all $\mathbf{x}, \mathbf{y} \in V$ and

$$f(a\mathbf{x}) = af(\mathbf{x})$$

   for all $\mathbf{x} \in V$ and all $a \in F$.

   Let $L(V, W)$ denote the set of all linear mappings from the vector space V to the vector space W. By defining addition of mappings along with scalar multiplications, show that the set $L(V, W)$ can be made into a vector space over the field F.

10. Which of the following form vector spaces over the given field?

   (a) The set of all real numbers of the form

$$a + b\sqrt{2} + c\sqrt[3]{3}$$

   where $a, b, c \in \mathbb{Q}$, the field of rational numbers.

   (b) The set of all polynomials of degree greater than five over a field F.

   (c) The set of all real functions f such that $f(x + 1) = f(x)$ over the field of real numbers.

   (d) The set of all polynomials with zero constant terms over a field F.

   (e) The set

$$\{0, x + 2, 2x + 4, 3x + 1, 4x + 3\}$$

   of polynomials in the variable x over the field $\mathbb{Z}_5$ of integers modulo 5.

11. Show that for any vectors $\mathbf{x}, \mathbf{y}$ in a vector space V over a field F and any scalar λ in the field F, the following hold.

   (a) $0\mathbf{x} = 0$ (here "0" is the scalar number 0);

   (b) $\lambda \mathbf{0} = \mathbf{0}$ (here "$\mathbf{0}$" is the zero vector);

   (c) $\lambda(\mathbf{x} - \mathbf{y}) = \lambda \mathbf{x} - \lambda \mathbf{y}$;

   (d) $-1\mathbf{x} = -\mathbf{x}$.

6.2 Basic properties of vector spaces

In this section we consider some of the basic properties of vector spaces. These properties include a discussion of linearly independent and linearly dependent vectors, sets of vectors that span a vector space, bases, and the dimension of a vector space. We provide numerous examples to illustrate these very important concepts.

Let $\mathbf{v}_1, \ldots, \mathbf{v}_n$ be a set of n vectors in a vector space V defined over a field F. We say that this set of vectors is **linearly dependent** if there are scalars $c_1, \ldots, c_n \in F$, not all zero, so that

$$c_1 \mathbf{v}_1 + \cdots + c_n \mathbf{v}_n = \mathbf{0}.$$

If the set is not linearly dependent, it is **linearly independent**. Equivalently, the set of vectors is linearly independent over the field F if, whenever

$$c_1 \mathbf{v}_1 + \cdots + c_n \mathbf{v}_n = \mathbf{0},$$

it must be the case that $c_1 = \cdots = c_n = 0$.

Example 6.9 *The vectors $(1,1)$ and $(4,5)$ are linearly independent over the field of real numbers. This is true because, if*

$$c_1(1,1) + c_2(4,5) = (0,0),$$

then we have the pair of simultaneous linear equations

$$c_1 + 4c_2 = 0$$
$$c_1 + 5c_2 = 0,$$

whose only solution is $c_1 = c_2 = 0$.

Remark 6.10 *In general, the vectors (a,b) and (c,d) are linearly independent over the real numbers, in fact over any field F, if $ad - bc \neq 0$.*

As another example we consider the following.

Example 6.11 *The vectors $(1,1,1), (1,2,1)$, and $(3,5,3)$ are linearly dependent over the field of real numbers. The reader should check this as in the above example by setting up a system of three simultaneous linear equations that has a non-zero solution. Hence the vectors are linearly dependent. One may also notice that the third vector is the sum of the first vector plus twice the second vector, so the three vectors as a set must be linearly dependent.*

Let V be a vector space over a field F. A vector

$$c_1\mathbf{v}_1 + \cdots + c_n\mathbf{v}_n$$

is a **linear combination** of the vectors $\mathbf{v}_1, \ldots, \mathbf{v}_n \in V$.

A set $\{\mathbf{v}_1, \ldots, \mathbf{v}_n\}$ of vectors in a vector space V is said to **span** the vector space V if any vector $\mathbf{v} \in V$ can be written in the form

$$\mathbf{v} = c_1\mathbf{v}_1 + \cdots + c_n\mathbf{v}_n$$

for some set of scalars $\{c_1, \ldots, c_n\}, c_i \in F$; i.e., if every vector in V can be written as a linear combination of the vectors $\mathbf{v}_1, \ldots, \mathbf{v}_n$.

In addition, a set of vectors forms a **basis** for the vector space V if it is both linearly independent and the set spans V. The **dimension** of a finite dimensional vector space V is the number of vectors in any basis (this number of vectors must be the same for any two bases; see page 44 of [14]).

Theorem 6.12 *Any two bases of a finite dimensional vector space over a field F contain the same number of vectors.*

In the above terminology, the sets $\mathbb{R}^2, \mathbb{R}^3$, and $\mathbb{R}^n$ $(n \geq 2)$ with the usual vector addition and scalar multiplication form vector spaces of dimensions 2, 3, and n over the field of real numbers.

In particular, bases can be given for each of these vector spaces as follows. The set of vectors

$$\{(1,0),(0,1)\}$$

forms a basis for $\mathbb{R}^2$ over the field of real numbers. Similarly the vectors

$$\{(1,0,0),(0,1,0),(0,0,1)\}$$

form a basis for $\mathbb{R}^3$ over the field of real numbers. And more generally, the set of vectors

$$\{(1,0,0,\ldots,0),(0,1,0,\ldots,0),\ldots,(0,0,0,\ldots 0,1)\}$$

forms a basis for $\mathbb{R}^n$ over the field of real numbers.

These bases are usually called **standard bases** for the vector spaces $\mathbb{R}^2, \mathbb{R}^3$, and $\mathbb{R}^n$ over the field of real numbers.

Actually, the above vectors also form a basis of the corresponding vector spaces $F^2, F^3, \ldots, F^n$ where F is any field, finite or infinite. The only difference is that the 1 in the vectors is the multiplicative identity of the field F, not the real number 1.

Example 6.13 *The set of all $n \times n$ matrices with real coefficients is a vector space of dimension n^2 over the field of real numbers. For a basis, the reader should consider the set of all $n \times n$ matrices where one entry is 1 and the remaining $n^2 - 1$ entries are 0. Here of course the 1 is moved through each of the n^2 cells of the matrix.*

Example 6.14 *The vector space V of all polynomials over a field is an infinite dimensional vector space. The reader should check that the infinite set of vectors*

$$\{1, x, x^2, \ldots, x^n, \ldots\}$$

forms a basis of this vector space V.

Similarly, the vector space of all polynomials of degree at most n over a field F at most n is of dimension $n + 1$. A basis is given by the set of polynomials

$$\{1, x, x^2, \ldots, x^n\}.$$

We now consider several small finite fields and indicate how they form vector spaces over subfields.

We first consider the field

$$\mathbb{F}_4 = \mathbb{F}_{2^2} = \{0, 1, \alpha, \alpha + 1\}$$

where α is a root of the irreducible polynomial $x^2 + x + 1$ over the field $\mathbb{Z}_2$ so that $\alpha^2 + \alpha + 1 = 0$, and hence $\alpha^2 = -\alpha - 1 = \alpha + 1$ since we are computing the coefficients modulo 2.

Consider the vectors (field elements) 1 and α. We claim that these form a basis over the field $\mathbb{Z}_2$.

Assume that $c_1 + c_2\alpha = 0$ for scalars $c_1, c_2 \in \mathbb{Z}_2$. By equating coefficients of these polynomials in the variable α, we immediately see that $c_1 = 0$ and hence $c_2 = 0$, so the vectors 1 and α are indeed linearly independent over the binary field $\mathbb{Z}_2$ of integers modulo 2.

The following calculations show that these vectors also span the field $\mathbb{F}_4 = \mathbb{F}_{2^2}$ as a vector space over the field $\mathbb{Z}_2$:

$$0 = 0(1) + 0(\alpha)$$

$$1 = 1(1) + 0(\alpha)$$

$$\alpha = 0(0) + 1(\alpha)$$

$$\alpha + 1 = 1(1) + 1(\alpha).$$

Thus, given an element $\beta \in \mathbb{F}_4$, it is clear that β can be written as a linear combination of the vectors 1 and α where the coefficients come from the base field $\mathbb{Z}_2$. A basis like $\{1, \alpha\}$ is called a **polynomial basis**.

In general, a basis

$$\{1, \alpha, \alpha^2, \ldots, \alpha^{n-1}\}$$

is a polynomial basis of the field $\mathbb{F}_{p^n}$ over the base field $\mathbb{Z}_p$. It is shown in

Section 1.5 of [25] that such a polynomial basis exists for the field $\mathbb{F}_{p^n}$ for any prime p and integer $n \geq 2$.

The reader should check that the vectors $\alpha, \alpha^2 = \alpha + 1$ also form a basis of the field $\mathbb{F}_4 = \mathbb{F}_{2^2}$ over the base field $\mathbb{Z}_2$. Such a basis is called a **normal basis**. As indicated below, it turns out that every finite field $\mathbb{F}_{p^n}$ has a normal basis

$$\{\alpha, \alpha^p, \ldots, \alpha^{p^{n-1}}\}$$

over the base field $\mathbb{Z}_p$; see Theorem 2.35 of [22].

Now consider the field $\mathbb{F}_8 = \mathbb{F}_{2^3}$ with 8 elements as a vector space over the base field $\mathbb{Z}_2$. As illustrated in Chapter 5, we can construct this field by considering the irreducible polynomial $p(x) = x^3 + x + 1$ over the field $\mathbb{Z}_2$ consisting of two elements. Let α be a root of $p(x)$ so that $\alpha^3 + \alpha + 1 = 0$ and hence $\alpha^3 = -\alpha - 1 = \alpha + 1$.

The reader should now check that the set of elements $\{1, \alpha, \alpha^2\}$ forms a polynomial basis of the field $\mathbb{F}_8$ over the field $\mathbb{Z}_2$. In addition, check that the set of elements

$$\{\alpha^3, (\alpha^3)^2, (\alpha^3)^{2^2}\} = \{\alpha^3, \alpha^6, \alpha^5\}$$

forms a normal basis over the field $\mathbb{Z}_2$.

The reader may ask, why is a normal basis useful in performing finite field arithmetic? The reason is that in doing finite field calculations, one often has to take p-th powers of an element, and for this kind of calculation, normal bases are very nice to have. We illustrate with the following example.

Example 6.15 *Consider the field $\mathbb{F}_{2^3}$ over the base field $\mathbb{Z}_2$ and assume that $\{\alpha, \alpha^2, \alpha^{2^2}\}$ is a normal basis. Assume that*

$$\beta = a_0\alpha + a_1\alpha^2 + a_2\alpha^{2^2}.$$

Then as the reader should check,

$$\beta^2 = a_2\alpha + a_0\alpha^2 + a_1\alpha^{2^2}$$

and

$$\beta^4 = (\beta^2)^2 = a_1\alpha + a_2\alpha^2 + a_0\alpha^{2^2}.$$

Note that the coefficients of β^2 are a_2, a_0, a_1, a simple shift to the right, of the coefficients a_0, a_1, a_2 of β. Similarly the coefficients a_1, a_2, a_0 of β^4 are a simple shift to the right of the coefficients a_2, a_0, a_1 of β^2.

Finally, we briefly discuss the most general situation where we consider the finite field $\mathbb{F}_{p^n}$ over the base field $\mathbb{Z}_p$ of integers modulo the prime number p.

Let $p(x)$ be an irreducible polynomial of degree n over the integers $\mathbb{Z}_p$ modulo p. As in the special cases above, let α be a root of $p(x)$. The reader should then check that the set of elements

$$\{1, \alpha, \alpha^2, \ldots, \alpha^{n-1}\}$$

forms a polynomial basis for the vector space $\mathbb{F}_{p^n}$ over the field $\mathbb{Z}_p$.

Here again, there is always a normal basis of the field $\mathbb{F}_{p^n}$ over the field $\mathbb{Z}_p$ (see Theorem 1.5.8 of [25]), although it is not so easy to write down the actual elements in such a general setting.

6.2 Exercises

1. Are the vectors $(2, 3)$ and $(4, 9)$ linearly independent or linearly dependent over the field of real numbers?

2. Are the matrices $A = \begin{pmatrix} 2 & 1 \\ -3 & 4 \end{pmatrix}$ and $B = \begin{pmatrix} 1 & -1 \\ 2 & -1 \end{pmatrix}$ linearly independent or linearly dependent over the field of real numbers?

3. The complex numbers form a vector space over the real numbers. What is the dimension of this vector space? Find a basis for this vector space.

4. Let $p \geq 5$ be a prime number. Over the field $\mathbb{Z}_p$ of integers modulo the prime p, are the vectors $(2, 4)$ and $(4, 3)$ linearly independent or linearly dependent?

5. Consider the three vectors $(2, 3, 4), (3, 4, 5)$, and $(4, 5, 6)$ over the field of real numbers. Are these vectors linearly independent?

6. Consider the three vectors $(2, 1, 3), (2, 2, 4)$, and $(3, 2, 1)$ over the real numbers. Are these vectors linearly independent?

7. If $\{\mathbf{u}, \mathbf{v}, \mathbf{w}\}$ is a set of linearly independent vectors in a vector space over a field F, show that the set $\{\mathbf{u}, \mathbf{u} + \mathbf{v}, \mathbf{u} + \mathbf{v} + \mathbf{w}\}$ is also linearly independent.

8. Assume that $\{\mathbf{u}_1, \mathbf{u}_2, \ldots, \mathbf{u}_n\}$ is a set of linearly dependent vectors over a field. Show that one of these vectors must be a linear combination of the others.

9. Show that two vectors in a vector space over a field F are linearly dependent over F if and only if one of them is a scalar multiple of the other.

10. Let F be a field. Let $n \geq 2$ be a positive integer. Show that the standard bases for each of the vector spaces $F^2, F^3, \ldots, F^n$ over the field F really do form bases, i.e., show that the sets of vectors are linearly independent over the field F and that they span the vector space.

11. Construct an infinite number of different bases for the vector space $\mathbb{R}^2$ over the field $\mathbb{R}$. Recall that since this vector space has dimension two, any basis must contain exactly two linearly independent vectors.

12. Let $\mathbf{x}$ and $\mathbf{y}$ be two vectors in a vector space V over the field F and let $a, b \in F$. Show that the vectors $\mathbf{x}$, $\mathbf{y}$ and $a\mathbf{x} + b\mathbf{y}$ form a linearly dependent set of vectors.

13. Let $\mathbf{x}, \mathbf{y}$, and $\mathbf{z}$ be vectors in a vector space V over a field F. Let $a, b \in F$. Show that the set $\{\mathbf{x}, \mathbf{y}, \mathbf{z}\}$ is a linearly dependent set if and only if the set

$$\{\mathbf{x} + a\mathbf{y} + b\mathbf{z}, \mathbf{y}, \mathbf{z}\}$$

is a linearly dependent set over the field F.

14. Let $\mathbf{x}$ and $\mathbf{y}$ be linearly independent vectors in a vector space over a field F. Let $a, b, c, d \in F$. Show that the vectors $a\mathbf{x} + b\mathbf{y}$ and $c\mathbf{x} + d\mathbf{y}$ are linearly independent over the field F if and only if $ad - bc \neq 0$.

15. Show that any set of $n+1$ vectors in a vector space of dimension n must be a set of linearly dependent vectors.

16. Let p be a prime and consider the field $\mathbb{Z}_p$ of integers modulo p. Assume that V is a vector space of dimension m over the field $\mathbb{Z}_p$. Determine the number of vectors in the vector space V.

17. Let

$$V = \left\{ \begin{pmatrix} a & b \\ b & c \end{pmatrix} \mid a, b, c \in \mathbb{Q} \right\}.$$

Show that V is a vector space over $\mathbb{Q}$. Find a basis for V over $\mathbb{Q}$. What is the dimension of the vector space V over the field of real numbers?

18. Show that every field is a vector space over itself. What is the dimension of this vector space? What is a basis?

19. If λ is a real number, show that the matrices $A = \begin{pmatrix} 1 & -2 \\ -3 & 7 \end{pmatrix}$ and $B = \begin{pmatrix} \lambda & -2\lambda \\ -3\lambda & 7\lambda \end{pmatrix}$ are linearly dependent over the field of real numbers.

20. If p is a prime, show that the vectors $(1, 2)$ and $(1, 1)$ are linearly independent over the finite field $\mathbb{Z}_p$.

21. Show that the vectors $(1, 2)$ and $(4, 3)$ are linearly independent over any finite field $\mathbb{Z}_p$ except over the field $\mathbb{Z}_5$ containing five elements.

22. Are the vectors $(1, 1, 1), (1, 2, 3)$, and $(1, 3, 4)$ linearly independent over the field $\mathbb{Z}_7$? Determine all prime fields $\mathbb{Z}_p$ for which these vectors are linearly independent over the field.

6.3 Subspaces

In this section we consider subsets of vector spaces, which themselves form vector spaces. Recall that we considered similar ideas when discussing subgroups of a group, subrings of a ring, and subfields of a field.

Let V be a vector space over a field F. A non-empty subset U of V is a **subspace** of V if U itself forms a vector space over the field F using the same operations as in the vector space V.

Thus, alternatively, if the subset U is closed with respect to addition and scalar multiplication, then U forms a subspace. Hence U is a subspace of V if and only if $a\mathbf{x} + b\mathbf{y} \in U$ for all $\mathbf{x}, \mathbf{y} \in U$ and all $a, b \in F$.

There are numerous examples of structures that form subspaces of various vector spaces.

For example, consider the vector space $V = \mathbb{R}^2$ over the field $\mathbb{R}$ of real numbers. Let U be the subset of V consisting of all vectors $(x, 0)$ with $x \in \mathbb{R}$. Since

$$(x_1, 0) + (x_2, 0) = (x_1 + x_2, 0)$$

and for any real number a

$$a(x, 0) = (ax, 0),$$

the subset U forms a subspace of the vector space V.

Similarly, if W consists of all vectors of the form $(0, y)$ with $y \in \mathbb{R}$, then W also forms a subspace of the vector space V.

We now give several examples of subspaces involving various kinds of matrices.

Let V denote the vector space of all $n \times n$ matrices over the field of real numbers. Let U denote the set of all $n \times n$ matrices with every element off the main diagonal equal to 0 (such matrices are called **diagonal matrices**). The reader should check that U is a subspace of the vector space V.

In addition, let W denote the set of all $n \times n$ matrices with entries in the field $\mathbb{R}$ of real numbers with every element off the main diagonal equal to zero and with a constant value on the diagonal; i.e., within a given matrix, each element on the main diagonal is the same. Such matrices are called **scalar matrices**. The reader should check that the scalar matrices form a subspace of the vector space of diagonal matrices. Scalar matrices also form a subspace of the vector space V consisting of all $n \times n$ matrices over the field of real numbers.

We refer to [2] for further details related to vector spaces and linear algebra.

6.3 Exercises

1. Let

$$S = \{(x, y, z) \mid x, y, z \in \mathbb{R}, x = 2y + 3z\}.$$

Show that the set S is a subspace of the vector space $\mathbb{R}^3$ over the field of real numbers. Find a set of vectors that forms a basis for the subspace S.

2. Assume that V is a vector space of dimension 5 over a field F. Further assume that U and W are subspaces of V, both of dimension 3. Show that $U \cap W \neq \{0\}$, i.e., show that $U \cap W$ must contain a non-zero vector.

3. Let V be the vector space $\mathbb{R}^3$ over the field of real numbers. Consider the set S of vectors (x, y, z) in V with the property that

$$x^2 + y^2 = z^2.$$

Is S a subspace of V? If so, what is its dimension?

4. Let U be a subspace of a vector space V and let $\mathbf{y} \notin U$. Show that the set

$$K = \{a\mathbf{y} + \mathbf{x} \mid \mathbf{x} \in U, a \in F\}$$

is a subspace of V.

5. Let $\mathbf{x}$ be a fixed vector in a vector space V over a field F. Show that $\{a\mathbf{x} \mid a \in F\}$ is a subspace of V.

6. (a) Show that the intersection of two subspaces of a vector space over a field forms a subspace of the vector space over the same field.

(b) Show that the intersection of any finite number of subspaces of a vector space over a field forms a subspace of the given vector space over the same field.

7. Show that any vector space V always has at least two subspaces.

8. If H and K are subspaces of a vector space V over a field F, show that the set

$$H + K = \{\mathbf{h} + \mathbf{k} \mid \mathbf{h} \in H \text{ and } \mathbf{k} \in K\}$$

is a subspace of V.

Chapter 7

Polynomials

In this chapter we discuss polynomials, mathematical objects with which we are very familiar from our studies in algebra and calculus. Polynomials occur in almost every mathematical setting. Even though the reader is likely to be familiar with polynomials, we begin by defining them rather carefully. We will consider polynomials defined over both rings and fields.

7.1   Basics

Assume that R is a commutative ring. The ring R might well be an integral domain or a field, but for now, we just assume that R is a commutative ring. A **polynomial** with coefficients in R is an expression of the form

$$p(x) = a_n x^n + a_{n-1} x^{n-1} + \cdots + a_1 x + a_0$$

where the coefficients $a_n, \ldots, a_0 \in R$, x is an indeterminate or a variable that can take on values from the ring R, and $n \geq 0$ is a non-negative integer.

A polynomial is **monic** if the coefficient of the highest power of x is 1, i.e., in our notation above, if $a_n = 1$.

A polynomial $p(x)$ has **degree** n if the coefficient a_n of the highest power of x is non-zero. Thus every polynomial except the zero polynomial (whose coefficients are all 0) has a degree that is at least 0. We define the degree of the zero polynomial to be -1.

Let

$$
\begin{aligned}
p(x) &= a_n x^n + \cdots + a_0 \\
q(x) &= b_m x^m + \cdots + b_0
\end{aligned}
\tag{7.1}
$$

be two polynomials with coefficients in the real numbers where we may assume that $m \leq n$. Then as polynomials, $p(x) = q(x)$ if the coefficients of each power of x are equal, i.e., if $a_i = b_i$, $i = 0, \ldots, m$ (and $a_{m+1} = \cdots = a_n = 0$).

Recall that two functions f and g are equal as functions on a ring R if $f(a) = g(a)$ for all $a \in R$. We note that if two polynomials p and q are equal as polynomials, then they are equal as functions; i.e., $p(a) = q(a)$ for all elements of $a \in R$.

One can however have two (actually an infinite number of) different polynomials that represent the same function. For example, over the field $\mathbb{Z}_2$ of integers modulo 2, each of the polynomials $x, x^2, x^3, \ldots$ are different as polynomials but they all yield the same function on $\mathbb{Z}_2$, namely the identity function that maps 0 to 0 and 1 to 1.

Usually our ring R will be fixed, but if one considers the same polynomial over two different rings, the behavior of the polynomial over the two rings can be very different! For example, consider the polynomial $f(x) = 2x$ over the real numbers. It is a linear polynomial. However, when considered as a polynomial over the field $\mathbb{Z}_2$ of integers modulo 2, we have

$$f(x) = 2x = 0x = 0$$

since $2 \equiv 0 \pmod 2$. Thus in this setting, the polynomial is not only a constant polynomial, it is the zero polynomial!

Given a ring R, we let $R[x]$ denote the set of all polynomials whose coefficients are in the ring R. We can now add and multiply polynomials in the usual way. To be more specific, the sum of the two polynomials in (7.1) of degrees $m \leq n$ is given by the polynomial

$$p(x) + q(x) = (a_n x^n + \cdots + a_0) + (b_m x^m + \cdots + b_0),$$

which upon simplification yields the polynomial

$$(a_0 + b_0) + (a_1 + b_1)x + \cdots + (a_m + b_m)x^m + a_{m+1}x^{m+1} + \cdots + a_n x^n.$$

In this notation, $a_0 + b_0$ is the sum of the elements a_0 and b_0 in the ring R, so the addition operation used is the additive operation for the ring R. Subtraction of polynomials works in a similar fashion.

We note that even though we are working over a commutative ring R, the way we add and multiply polynomials over the ring is the same as we do it over the real numbers. For multiplication, we have the product $p(x)q(x)$, which is given by

$$a_0 b_0 + (a_0 b_1 + a_1 b_0)x + \cdots + \left(\sum_{i+j=k} a_i b_j \right) x^k + \cdots + a_n b_m x^{n+m}.$$

We note that if the ring R is an integral domain, then the degree of the product of two polynomials is the sum of the degrees of the individual polynomials.

The reader should check that if R is a commutative ring, then so is $R[x]$, the set of all polynomials whose coefficients lie in the ring R. In fact, if the ring R is an integral domain, then $R[x]$ is also an integral domain.

We also mention that the ring R itself can be viewed as a subset of the polynomial ring $R[x]$ by viewing an element $a \in R$ as a polynomial of degree 0 if $a \neq 0$, and of degree -1 if $a = 0$.

The reader will recall the Division Algorithm for integers from Theorem 1.2. It turns out that there is a polynomial version of the Division Algorithm that holds for polynomials over any field. In fact, the proof of the polynomial version is very similar to that given for the integer version, so we will omit this proof and simply state the algorithm:

Theorem 7.1 *(Division Algorithm) Let F be a field. Let f, g be two polynomials over F where $f \neq 0$. Then there are unique polynomials q and r with the degree of r less than the degree of f so that $g = fq + r$.*

The polynomials q and r are often called the **quotient** and **remainder** polynomials.

We now provide an example to illustrate the Division Algorithm over the field $\mathbb{Z}_2$ of integers modulo 2.

Example 7.2 *Using the Division Algorithm, if we divide $g(x) = x^5 + x^2$ by $f(x) = x^2 + 1$, we obtain*

$$x^5 + x^2 = (x^2 + 1)(x^3 + x + 1) + x + 1$$

so that $q(x) = x^3 + x + 1$ and $r(x) = x + 1$.

Next we state the Euclidean Algorithm for finding the greatest common divisor $\gcd(f, g)$ of two polynomials f and g over a field F. The proof repeatedly uses Theorem 7.1. Since the details of the proof of this result are very similar to those given in Chapter 1 for the integer version, we will omit the proof and instead illustrate the algorithm with an example.

Theorem 7.3 *(Euclidean Algorithm) Let f and g be polynomials over a field F. If f divides g then the $\gcd(f, g) = f$. Otherwise, there are non-zero polynomials $r_1, \ldots, r_n$ of decreasing degrees so that*

$$
\begin{aligned}
g &= fq_1 + r_1 \\
f &= r_1 q_2 + r_2 \\
r_1 &= r_2 q_3 + r_3 \\
&\vdots \\
r_{n-2} &= r_{n-1} q_n + r_n \\
r_{n-1} &= r_n q_{n+1}.
\end{aligned}
$$

Then $\gcd(f, g) = r_n$.

Example 7.4 *By way of illustration, we use the Euclidean Algorithm to find the greatest common divisor of the polynomials $x^4 + x^3 + 1$ and $x^2 + x + 1$ over the field $\mathbb{Z}_2$ of integers modulo 2. By repeated use of the Division Algorithm, we have*

$$
\begin{aligned}
x^4 + x^3 + 1 &= (x^2 + x + 1)(x^2 + 1) + x \\
x^2 + x + 1 &= x(x) + x + 1 \\
x &= (x + 1)1 + 1 \\
x + 1 &= 1(x + 1) + 0.
\end{aligned}
$$

Thus, $\gcd(x^4 + x^3 + 1, x^2 + x + 1) = 1$. Therefore the polynomials are relatively prime.

Using the Division Algorithm for polynomials from Theorem 7.1, one can prove the following result concerning polynomials over a field.

Theorem 7.5 *If f is a polynomial over a field F and $a \in F$, then $f(a)$ is the remainder when $f(x)$ is divided by $x - a$.*

Proof: By the Division Algorithm for polynomials,

$$f(x) = (x - a)q(x) + r(x)$$

where the degree of $r(x)$ is less than the degree of $x - a$ (which is 1). The result follows. ∎

Our next theorem provides an important result concerning polynomials over a field and their roots. It follows from Theorem 7.5.

Theorem 7.6 *Let f be a polynomial whose coefficients lie in a field F, and let a be an element in F. Then $f(a) = 0$ if and only if $x - a$ divides $f(x)$.*

We close this section with a very important result concerning the number of roots of a polynomial of degree n over a field.

Theorem 7.7 *Let f be a non-zero polynomial of degree $n \geq 1$ whose coefficients lie in a field F. Then f has at most n roots in the field F.*

Proof: Our proof will be by induction on the degree n.

If the polynomial f has degree 1 then it is a linear polynomial, say $ax + b$ with $a \neq 0$, so that $x = -\frac{b}{a}$ is a root in the field F.

Assume now that f is a polynomial of degree $n > 1$. If a is a root of $f(x)$ so that $f(a) = 0$, then

$$f(x) = (x - a)g(x).$$

Let c be a root of $f(x)$ with $c \neq a$. Then $f(c) = 0$ and so c is a root of the polynomial $g(x)$ since $c \neq a$ and the field F is an integral domain.

Thus the only roots of $f(x)$ are a and the roots c of $g(x)$. By the induction hypothesis, $g(x)$ has degree $n - 1$, so $g(x)$ has at most $n - 1$ roots, and hence, $f(x)$ has at most n roots.

If there is no root c of $f(x)$ with $c \neq a$, then

$$f(x) = (x - a)g(x)$$

and the polynomial $g(x)$ has no root in the field F. But then $f(x)$ has only one root in F and $1 \leq n$, so the proof is complete. ∎

7.1 Exercises

1. Let $f(x) = x^3 - 5x^2 + 2x - 4$ be a polynomial over the field $\mathbb{Z}_5$ of integers modulo 5. Calculate $f(a)$ for each $0 \leq a \leq 4$.

2. Find the quotient $q(x)$ and the remainder $r(x)$ when $x^4 - x^2 + 1$ is divided by $x + 2$ as polynomials over the field $\mathbb{Z}_3$ of three elements.

3. Do the polynomials $x^4 + 1$ and $x^2 + x + 1$ have a common factor other than 1 over the field $\mathbb{Z}_2$?

4. Find all solutions to the equation $x^2 - 3x + 2 = 0$ in the ring $\mathbb{Z}_6$. Note that there are more than two solutions. Why doesn't this example contradict Theorem 7.1?

5. Find two different polynomials over $\mathbb{Z}_6$ that give the same function on $\mathbb{Z}_6$.

7.2   Unique factorization

We begin by defining what is meant by an irreducible polynomial, i.e., a polynomial that cannot be factored into polynomials with smaller degrees. More specifically, a polynomial $f(x)$ is **irreducible** if there are no polynomials $g(x)$ and $h(x)$ of positive degrees over the field F so that $f(x) = g(x)h(x)$.

The reader should check that with this definition, polynomials of degree two or three are irreducible if and only if they do not have a root in the field. Be careful, this result for polynomials of degrees two and three does not hold for polynomials of degree four or higher!

Theorem 7.8 *Let F be a field. A polynomial f over F of degree at least one over F is irreducible, or it factors into a product of irreducible polynomials.*

We will omit a proof of this result as the proof is very similar to the proof given for the fact that a positive integer $n \geq 2$ is a prime or that it factors into a product of primes, as discussed in Chapter 1.

In fact, as with the number–theoretic result that the factorization into primes is unique except for the order of the factors, the same uniqueness of factorization holds for polynomials over any field. We can combine like factors and thus we may state this result as follows:

Theorem 7.9 *Let F be a field and let $f \in F[x]$. Then, except for the order of the factors, f may be uniquely written as*

$$f = f_1^{a_1} \cdots f_r^{a_r}$$

where each $f_i, i = 1, \ldots, r$, is irreducible and each $a_i \geq 1, i = 1, \ldots, r$.

As several illustrations of this important result, we mention the following:

Example 7.10 *We have $(x^2 - 16)^3 = (x - 4)^3(x + 4)^3$ over the field of real numbers.*

Example 7.11 *Over the real numbers $x^2 + 1$ is irreducible, but over the complex numbers it factors into irreducibles as $(x - i)(x + i)$.*

Example 7.12 *Over the field $\mathbb{Z}_2$ of integers modulo 2, we have the factorization*

$$x^4 - 1 = (x^2 - 1)(x^2 + 1) = (x - 1)(x + 1)(x + 1)(x + 1) = (x + 1)^4,$$

since $-1 \equiv 1 \pmod{2}$.

7.2 Exercises

1. Factor the polynomial $x^3 - 1$ into a product of irreducibles over the fields $\mathbb{Z}_2$ and $\mathbb{Z}_3$.

2. Factor the polynomial $x^3 + x^2 + x + 1$ over the fields $\mathbb{Z}_2$ and $\mathbb{Z}_3$.

3. Factor the polynomial $x^3 - x^2 + x + 1$ over the field $\mathbb{Z}_3$.

4. List all irreducible polynomials of degrees at most four over the binary field $\mathbb{Z}_2$.

5. Find all roots of the polynomial $x^4 + x^2 + 1$ over the fields $\mathbb{Z}_2$ and $\mathbb{Z}_3$.

6. Show that over the field $\mathbb{Z}_5$, the polynomial $(x - 2)^5 = x^5 - 2$.

7. Show over a field of characteristic p prime that

$$(a+b)^p = a^p + b^p$$

for any elements a and b in the field $\mathbb{Z}_p$. (Recall the characteristic of a field must be 0 or a prime p.)

8. Show that over a field of characteristic p,

$$(a+b)^{p^n} = a^{p^n} + b^{p^n}$$

for any elements a and b in the field and any non-negative integer n.

7.3 Polynomials over the real and complex numbers

In this section we briefly discuss several results concerning the irreducibility of polynomials over the field of real numbers and over the field of complex numbers. The reader will detect some significant differences in these two settings.

Clearly any linear polynomial $ax + b$ with $a \neq 0$ is irreducible over both the real numbers and over the complex numbers. We recall from our studies of algebra that when considering a quadratic polynomial $ax^2 + bx + c$, the value $b^2 - 4ac$ plays an important role. This value is called the **discriminant** of the polynomial.

Theorem 7.13 *A quadratic polynomial $ax^2 + bx + c \in \mathbb{R}[x]$ is irreducible over the field $\mathbb{R}$ of real numbers if and only if the discriminant $b^2 - 4ac$ is negative.*

We can say even more. Given a quadratic polynomial $ax^2 + bx + c$ with real coefficients, the discriminant can be used to indicate whether the polynomial has no real roots (in this case $b^2 - 4ac < 0$), one multiple (double) real root (in this case $b^2 - 4ac = 0$), or two distinct real roots (in this case $b^2 - 4ac > 0$). We will illustrate each of these ideas with the following examples.

Consider the polynomial $x^2 + x + 1$ whose discriminant is $b^2 - 4ac = -3$. The polynomial does not have any real roots, but by the quadratic formula, it does have two complex roots $\frac{-1 \pm \sqrt{3}i}{2}$ (each of which is the complex conjugate of the other). The reader should check that these roots really do satisfy the polynomial equation $x^2 + x + 1 = 0$.

The polynomial $x^2 + 2x + 1 = (x+1)^2$ has discriminant $b^2 - 4ac = 0$ and it has a double root $x = 1$ of multiplicity two.

Finally, consider the polynomial $x^2 + 3x + 2$ so that $b^2 - 4ac = 1$. This polynomial has two real and distinct roots $x = 1$ and $x = 2$.

The above discussion shows that there are irreducible polynomials of degrees one and two over the real numbers. This leads to the following question: Can you find an irreducible polynomial with coefficients in the real numbers whose degree is greater than two?

While beyond the scope of our text, it turns out that the following is true:

Theorem 7.14 *There are no polynomials over the real numbers of degree greater than two that are irreducible.*

The above theorem explains why, in the study of partial fractions, one only needs to consider denominators that are powers of linear or quadratic polynomials. Recall that partial fractions can be quite helpful in integration problems to break an integral into several integrals that are simpler to evaluate.

In striking contrast, Gauss (1777–1855), who was one of the most profilic mathematicians of the early 19th century, proved the following very important result in 1801 in his famous text, *Disquisitiones Arithmeticae*, while working over the field $\mathbb{C}$ of complex numbers. In [5] the authors provide a reprinted English version of the book *Disquisitiones Arithmeticae*.

Theorem 7.15 *(Fundamental Theorem of Algebra) Every polynomial $f(x) \in C[x]$ of degree greater than one has a root in the field C of complex numbers.*

Using this theorem, we can quickly see that the following result must also hold.

Theorem 7.16 *The only irreducible polynomials over the complex numbers are of degree one.*

You should convince yourself why this result holds.

7.3 Exercises

1. Factor the polynomial $x^2 + 5$ over the field of complex numbers.

2. Factor the polynomial $x^4 + 1$ over the field of real numbers.

3. Factor the polynomial $x^4 - 1$ over the field of complex numbers.

7.4   Root formulas

In this section, we discuss a few results concerning the problem of solving polynomial equations motivated by the quadratic formula.

Given a linear equation

$$ax + b = 0$$

over the real numbers with $a \neq 0$, we can find a formula for the solution; namely $x = -\frac{b}{a}$.

Similarly for a quadratic equation

$$ax^2 + bx + c = 0,$$

the quadratic formula can be employed to show that we have two solutions,

$$x = \frac{-b \pm \sqrt{b^2 - 4ac}}{2a}. \tag{7.2}$$

It is important to note that the quadratic formula was, in essence, known to Babylonian mathematicians several thousand years ago.

It is instructive to see just how the two roots in the quadratic formula (7.2) arise. The primary tool that is needed is often known as "completing the square."

We begin with the quadratic equation

$$ax^2 + bx + c = 0$$

where $a \neq 0$. We then factor the constant a from the first two terms of the left-hand side of the equation.

$$a \left(x^2 + \frac{b}{a}x \right) + c = 0.$$

Next, we complete the square by adding a constant term within the parentheses on the left-hand side of the equation so that the terms within the parentheses form a square. In this instance, the amount to be added is $\left(\frac{b}{2a} \right)^2$. Of course, if we add within the parentheses, then we must add the corresponding amount to the right-hand side of the equation (in order for the equation to remain true). Note that the amount to add to the right-hand side is $a \left(\frac{b}{2a} \right)^2$ or $\frac{b^2}{4a}$ because of the presence of the factor a, which is outside the parentheses. This now yields the following equation:

$$a \left(x^2 + \frac{b}{a}x + \left(\frac{b}{2a} \right)^2 \right) + c = \frac{b^2}{4a}.$$

We now simplify this equation via several steps of algebra to obtain the following:

$$a \left(x^2 + \frac{b}{a}x + \left(\frac{b}{2a} \right)^2 \right) + c = \frac{b^2}{4a}$$

$$a\left(x+\frac{b}{2a}\right)^2 + c = \frac{b^2}{4a}$$

$$a\left(x+\frac{b}{2a}\right)^2 = \frac{b^2}{4a} - c$$

$$a\left(x+\frac{b}{2a}\right)^2 = \frac{b^2 - 4ac}{4a}$$

$$\left(x+\frac{b}{2a}\right)^2 = \frac{b^2 - 4ac}{4a^2}$$

$$x+\frac{b}{2a} = \pm\sqrt{\frac{b^2 - 4ac}{4a^2}}$$

$$x+\frac{b}{2a} = \pm\frac{\sqrt{b^2 - 4ac}}{2a}$$

$$x = -\frac{b}{2a} \pm \frac{\sqrt{b^2 - 4ac}}{2a}$$

$$x = \frac{-b \pm \sqrt{b^2 - 4ac}}{2a}.$$

Note that for equations of degrees one and two, we can always find a solution by just using additions, subtractions, multiplications, and divisions as well as radicals. Using just these operations to solve a polynomial equation is known as **solving the equation by radicals**.

What happens for polynomials of degree three? Is there also a formula involving radicals? The answer is yes.

The cubic equation was first solved by the Italian mathematician Scipione del Ferro (1465–1526), although he was reluctant to share these results widely and never published them. A few decades later, formulas for cubic equations were again discovered by two additional Italian mathematicians, Nicolo Fontana (1500–1557), better known as Tartaglia, and Girolamo Cardano (1501–1576). Sadly, the relationship between Tartaglia and Cardano became quite antagonistic, especially after Cardano published the solution of the cubic equation in his *Ars Magna* in 1545 (at the protestation of Tartaglia).

For the equation

$$ax^3 + bx^2 + cx + d = 0,$$

a (much more complicated) formula can be obtained using the process of solving by radicals. A formula for a solution is given below.

We will see that this can indeed be done, but after looking at the formula, we think you will agree that we really don't want to pursue this idea of solving cubic equations by radicals as a method to really solve equations.

The solution x of

$$ax^3 + bx^2 + cx + d = 0$$

is given by

$$\sqrt[3]{\left(\frac{-b^3}{27a^3} + \frac{bc}{6a^2} - \frac{d}{2a}\right) + \sqrt{\left(\frac{-b^3}{27a^3} + \frac{bc}{6a^2} - \frac{d}{2a}\right)^2 + \left(\frac{c}{3a} - \frac{b^2}{9a^2}\right)^3}}$$

$$+ \sqrt[3]{\left(\frac{-b^3}{27a^3} + \frac{bc}{6a^2} - \frac{d}{2a}\right) - \sqrt{\left(\frac{-b^3}{27a^3} + \frac{bc}{6a^2} - \frac{d}{2a}\right)^2 + \left(\frac{c}{3a} - \frac{b^2}{9a^2}\right)^3}}$$

$$- \frac{b}{3a}.$$

A formula such as this one was known to Cardano in the sixteenth century.

Assume that one is trying to solve a cubic equation

$$ax^3 + bx^2 + cx + d = 0,$$

where we can assume that $a \neq 0$.

We first note that if we replace x by the substitution $x - \frac{3b}{a}$, we obtain an equation without a quadratic term. Further, one could also divide by a to yield an equation of the form

$$x^3 + \alpha x + \beta = 0.$$

For example, assume that we want to solve the cubic equation

$$x^3 + 6x - 20 = 0.$$

Here $a = 1$ and $b = 0$, greatly simplifying the calculation in the above radicals. After some arithmetic we find that a solution x is given by

$$\sqrt[3]{10 + \sqrt{108}} + \sqrt[3]{10 - \sqrt{108}}.$$

It is not obvious algebraically that $x = 2$, but by use of a calculator, we see that x is very likely equal to 2. How do we prove that this is indeed the case?

If we divide $x^3 + 6x - 20$ by $x - 2$ using the Division Algorithm for polynomials, we see that

$$x^3 + 6x - 20 = (x - 2)(x^2 + 2x + 10).$$

Hence $x = 2$ is indeed a root of the cubic equation.

Now to obtain the other two roots, we apply the quadratic formula to the polynomial $x^2 + 2x + 10$ to obtain $x_2 = -1 + 3i$ and $x_3 = -1 - 3i$, and we are done.

The alert reader may have noticed that the integer root $x = 2$ divides the constant term -20 of the polynomial. This is no coincidence, as is proved in the following result.

Theorem 7.17 *If x is an integer root of the integer polynomial equation*

$$a_n x^n + a_{n-1} x^{n-1} + \cdots + a_1 x + a_0 = 0,$$

then x must divide the constant term a_0.

Proof: If x is an integer root of the above polynomial equation, then

$$a_n x^n + a_{n-1} x^{n-1} + \cdots + a_1 x = -a_0 = 0,$$

or

$$x(a_n x^{n-1} + a_{n-1} x^{n-2} + \cdots + a_1) = -a_0,$$

and hence x must divide the constant term a_0. ∎

Thus had we known this theorem before we tried using the complicated formula involving cubic and quadratic roots, we could have saved ourselves a lot of algebraic work.

Given a polynomial $p(x)$ with integer coefficients, in order to check whether the polynomial has an integer solution, one only needs to check for each divisor d of the constant terms a_0, if $p(\pm d) = 0$. Of course if the constant term a_0 has many divisors, this can involve considerable time and effort. On the other hand, if no integer solution is found, this does not mean that the polynomial equation does not have a solution, as it may well have a rational or irrational solution, or a complex solution. As an illustration of a polynomial equation that does not have any integer, rational, or real (not even irrational) solution, consider the very simple equation $x^2 + 1 = 0$ whose roots are $\pm i$, which are, of course, complex numbers.

One might now ask whether we can we say anything about when there might be a rational solution to our polynomial equation.

Let's try to generalize the previous integer argument to rational numbers. We now prove the following.

Theorem 7.18 *If $x = \frac{\alpha}{\beta}, \beta \neq 0$ is a rational solution of the polynomial equation*

$$a_n x^n + a_{n-1} x^{n-1} + \cdots + a_1 x + a_0 = 0,$$

where the coefficients $a_n, \ldots, a_0$ are integers, then the numerator α must divide the constant term a_0 of the polynomial, and the denominator β must divide the leading coefficent a_n of the polynomial.

Proof: Let $x = \frac{\alpha}{\beta}$ be a solution of the given polynomial equation. We first put everything on the left-hand side over the common denominator β^n. Then the integer α must divide each of the terms in the numerators along with 0, and hence α must divide the constant term a_0. A similar argument shows that β must divide the leading coefficient a_n. ∎

This is great, we can check for integer and rational solutions of any polynomial equation! However, as indicated in the discussion regarding integer solutions, a similar caveat exists here for rational solutions as well; namely that if the leading coefficient a_n and the constant term a_0 have lots of divisors, we will have lots of checking to do before we can decide if there are any rational solutions.

As an illustration of a check for rational solutions, consider the polynomial equation

$$8x^2 - 14x + 3 = 0,$$

whose roots, by the quadratic formula, turn out to be $x = \frac{1}{4}$ and $x = \frac{3}{2}$.

Notice that the numerators of both roots divide the constant term 3, and the denominators 2 and 4 both divide the leading coefficient, which is 8, as they must by the above theorem.

The reader may have wondered why we didn't just use the quadratic formula to obtain these roots; the point is that we wanted to carefully illustrate the use of this theorem with a small, easily understood example.

It turns out that every polynomial equation of degree four over the reals can also be solved by radicals, although as one might surmise, the formula is even more complicated than for the degree-three case. A formula obtained by radicals for the degree-four polynomial equation was first obtained by Lodovico Ferrari (1522–1565), Cardano's student. The solution of the degree-four case, or quartic case, was also published by Cardano in *Ars Magna* in 1545.

For many years, mathematicians thought that this process continued, namely that a polynomial equation of any degree could always be solved by radicals. It was believed that the formulas would simply become more and more complicated as the degrees got larger and larger. That belief changed with the groundbreaking work of two young mathematicians.

As indicated in [10], the Norwegian mathematician Niels Abel (1802–1829) showed that for degree $n = 5$, there are polynomial equations that cannot be solved by radicals. Unfortunately, Abel's health was poor throughout his brief life, and he died in his native Norway at the age of 26. Later, the French mathematician Évariste Galois (1811–1832) proved that for any degree $n \geq 5$, there are polynomial equations of degree n that cannot be solved by radicals. Unfortunately, Galois' life was cut short during a duel in which he was mortally wounded.

It turns out that some very high-powered group theory was used to prove that for any degree $n \geq 5$, there are polynomial equations of degree n that cannot be solved by radicals. In particular, Galois proved that a polynomial equation can be solved by radicals if and only if the "Galois group" of the equation is "solvable" as a group. Unfortunately, the topics of solvable groups and Galois groups are beyond the scope of this text.

The result of Galois came as quite a surprise to the mathematical community. The work of Galois helped to develop the abstract theory of groups; see [10] for a very readable discussion of these topics.

While discussing equations that can be solved by radicals, it may be of interest to consider a non-trivial class of polynomial equations (not just polynomial equations like $x^n = 2$), which can be solved by radicals. Such a class of polynomials is given by what we earlier called **Dickson polynomials of the**

first kind, which are named for the American mathematician L. E. Dickson (1874 – 1954). Dickson, who is well-known for his publication [7] of the three-volume *History of the Theory of Numbers* (1919 – 1923), spent most of his influential career at the University of Chicago. He served as the president of the American Mathematical Society during the 1917 – 1918 timeframe, and received the Cole Prize for Algebra in 1928.

Dickson polynomials of the first kind of degree n with parameter a are defined by

$$D_n(x, a) = \sum_{i=0}^{\lfloor n/2 \rfloor} \frac{n}{n-i} \binom{n-i}{i} (-a)^i x^{n-2i},$$

where a is a real parameter and $\binom{A}{B}$ is the usual binomial coefficient, which, for nonnegative integers A and B, is given by

$$\binom{A}{B} = \frac{A!}{B!(A-B)!}.$$

The first few Dickson polynomials are

$$
\begin{aligned}
D_0(x, a) &= 2, \\
D_1(x, a) &= x, \\
D_2(x, a) &= x^2 - 2a, \\
D_3(x, a) &= x^3 - 3ax, \\
D_4(x, a) &= x^4 - 4ax^2 + 2a^2.
\end{aligned}
$$

Dickson polynomials are related to the Chebyshev polynomials [20, Chapter 7] and satisfy the second-order recurrence

$$D_{n+2}(x, a) = x D_{n+1}(x, a) - a D_n(x, a), \qquad n \geq 0 \qquad (7.3)$$

with $D_0(x, a) = 2$ and $D_1(x, a) = x$.

Note that if the parameter $a = 0$, the Dickson polynomial

$$D_n(x, 0) = x^n$$

is a simple power polynomial. While the Dickson polynomials may appear to be quite complicated, we note that they have a very nice functional equation to help simplify arithmetic calculations. We refer to [20] for many algebraic and number-theoretic properties of Dickson polynomials. We also refer to [19] for a proof that any Dickson polynomial equation of degree n can be solved by radicals.

7.4 Exercises

1. Find all roots of the polynomial $x^3 - 2x^2 - x + 2$ over the field $\mathbb{Z}_5$.

2. Find all roots of the polynomial from the preceding exercise over the fields $\mathbb{Z}_2$ and $\mathbb{Z}_7$.

3. Use the recurrence relation (7.3) for Dickson polynomials to calculate the Dickson polynomials $D_n(x, a)$ for $5 \leq n \leq 8$.

4. For $n \leq 5$, factor the Dickson polynomials $D_n(x, 1)$ into irreducibles over the field $\mathbb{Z}_2$.

5. Find all roots of the quadratic equation

$$x^2 - x + 1 = 0$$

over the field of real numbers.

6. Find all roots of the quadratic equation

$$x^2 - x + 1 = 0$$

over the field of complex numbers.

7. Find all roots of the cubic polynomial $x^3 - 7x + 6$ over the real numbers.

8. Find all roots of the cubic equation $x^3 + 4 = 0$ over the real numbers.

9. Find all roots of the polynomial equation $x^3 - 4x^2 - 9x + 36 = 0$ over the field of real numbers.

10. Factor the polynomial $x^3 - 6x^2 + 11x - 6$ over the field of real numbers.

11. Factor the polynomial $x^3 - 6x^2 + 11x - 6$ over the field $\mathbb{Z}_2$.

12. Factor the polynomial $x^3 - 6x^2 + 11x - 6$ over the field $\mathbb{Z}_3$.

13. Find all roots of the polynomial $x^4 - 1$ over the field of complex numbers.

14. Does the polynomial $3x^3 + 4x + 6$ have an integer root?

15. Does the polynomial $x^7 - 8x + 2$ have an integer root?

16. Find all roots of the polynomial $x^3 - 6x^2 + 11x - 6$ over the real numbers.

17. Does the polynomial $x^4 - 2x^2 + 2$ have any rational roots?

18. Does the polynomial $x^3 - 3x^2 + 4$ have any rational roots?

Chapter 8

Linear Codes

In Section 6 of Chapter 1, we briefly discussed a method to securely send messages from one point, A to another point, B. We provided a brief discussion of one such method, namely the RSA cryptosystem, which is based upon Euler's theorem (Theorem 1.30). The main focus there was to securely protect our message from unauthorized listeners.

Now we consider a different example of modern communications; namely, how do we **accurately** send a message from A to B? Suppose we have a file that contains financial data that A wants to accurately send to B (for example, a computer file to be sent from one computer to another). How can this be accomplished; i.e., how can we minimize the chance of errors being introduced during the transmission process?

One answer to this question involves a very practical application of Lagrange's Theorem (Theorem 2.13) from the theory of finite groups. Indeed, the topic of this chapter, linear codes, was developed primarily to answer this question.

Coding theory is the study of methods for the efficient and accurate transfer of information from one place to another. The theory has been developed for such diverse applications as the minimization of noise from compact disk recordings, the transfer of financial information across telephone lines, data transfer from one computer to another or from memory to the central processor, and for information transmission from a distant source such as a weather or communication satellite. For example, on November 1, 1980, Voyager 1 sent to the earth the first high-resolution pictures of Saturn and its moons using an error-correcting code. Voyager 2 sent even better images of Saturn in August 1981.

The study of such methods and systems is called **algebraic coding theory** because various kinds of algebra are often used. We note that, contrary to cryptographic systems discussed in Section 6 of Chapter 1, we are not worried about people listening in; in algebraic coding theory we just want to accurately move information from one place to another. Such codes are also called **error-correcting** codes. As will be seen shortly, one can indeed correct a limited number of errors by implementing such codes.

How do we go about sending the message "We love math!" to someone? First, as in cryptography, we must convert our message to numbers. Here we want to convert our message to a binary string; i.e., to a string of 0s and 1s so that the parts of the string can be viewed as elements of the binary field $\mathbb{Z}_2$ of integers modulo 2.

We now, very informally, consider a small example to illustrate the basic ideas from algebraic error-correcting coding theory. Assume that we have four binary messages, say $00, 01, 10$, and 11. When transmitting binary digits, it is of course possible that a 0 will be switched to a 1 or a 1 will be switched to a 0. For example, using a telephone line, there may well be background noise so that it becomes difficult, if not impossible, to be able to accurately send and receive the correct digits.

In order to enable us to correct errors, we don't just send the messages above. We first, very carefully, add some extra binary digits, called **parity check digits** or just **check digits**, to our message. (Many real–world examples of such "check digits" exist. For example, the last digit of the UPC codes found on products at your local store is a check digit, as is the last digit of the ISBN numbers on various books. Even the Vehicle Identification Number (or VIN) on your automobile employs a check digit.)

As will be seen shortly, we can encode our four messages to become four **codewords** each of **length** five, i.e., now with five binary digits. The reader will later see that one such method of encoding our messages is the following: The message 00 becomes 00000; 01 becomes 01011; 10 becomes 10110; and 11 becomes 11101. Hence our code C is given by

$$C = \{00000, 01011, 10110, 11101\}.$$

You might ask, "Why do we choose these extra digits in this particular way?" Well, notice that the sum modulo 2 of any two codewords is another codeword. Such a code is said to be **linear**. We should also note that there are $2^5 = 32$ binary vectors of length five and they form a commutative group G under the operation of addition of binary vectors where we add modulo 2 in each of the five coordinates.

In addition, there are $2^2 = 4$ codewords and they form a subgroup H of order four of the group G. This subgroup H is our linear code C. We note from Lagrange's Theorem (Theorem 2.13) that 4 divides 32, so our coding ideas fit nicely into the group theory studied in Chapter 2.

Note that in our example code C above, each non-zero codeword has **weight** three, i.e., each non-zero codeword has three non-zero coordinates. Hence the minimum weight of the non-zero codewords in the code is 3.

It will later be seen that if the minimum weight of the various non-zero codewords is at least $2t + 1$, then the code can correct t errors. Hence in our example, $3 = 2(1) + 1$, so our code C can correct $t = 1$ error. Thus we will be able to correct any single error that occurs, i.e., we can correct any single switch of a 0 to a 1 or any single switch of a 1 to a 0 in any codeword that occurs during the transmission process.

When we receive a vector, how do we decode it to obtain the original codeword that was sent? Assume that we have received the vector 11111. Notice that this vector differs from the codeword 00000 in five places; it differs from the codewords 01011 and 10110 in exactly two places, and it differs from the codeword 11101 in one place.

Since we know that our code can correct one but not more than one error, we decode the received vector 11111 to the codeword 11101. Later we will have a much better and more systematic algorithm for decoding the vectors we receive. In particular, we will not have to compare our received vector to each of the codewords in the code. But for now, the example suffices to illustrate that we can indeed correct some limited number of errors that may have arisen in the transmission process.

The key point to be taken from this discussion is the way in which we encoded our messages $00, 01, 10$, and 11 so that we can correct at least one error in our received messages.

8.1   Basics

We begin by defining what is called a linear **error-correcting** code, which because of the use of algebraic operations, is often called an **algebraic** code. We will make considerable use of some of our earlier group theoretic ideas, especially those related to subgroups of finite groups and Lagrange's Theorem.

Linear codes can be considered over any finite field $\mathbb{F}_q$, or even over more general finite rings; in particular, they can be considered over the field $\mathbb{Z}_p$ where p is any prime. However the development of coding theory using the field of p elements, or even more generally q elements, is basically the same as that using just the binary field. Thus, we will restrict our attention to the binary case. See Chapter 3 of [25] for further details of algebraic codes constructed over more general finite fields. Also see [14] and [29] for very readable texts on algebraic coding theory.

As we move ahead, we remind the reader that we are working over $\mathbb{Z}_2$, the binary field of integers modulo 2.

We first fix a positive integer n, which will be the **length** of our code.

Consider the set $\mathbb{Z}_2^n$ of all binary n-tuples over the binary field $\mathbb{Z}_2$. This set forms a commutative group G under addition, with the operation defined on two n-tuples by

$$(a_1, \ldots, a_n) + (b_1, \ldots, b_n) = (a_1 + b_1 \pmod 2, \ldots, a_n + b_n \pmod 2).$$

A **linear code** C is a subgroup C of the additive group $G = \mathbb{Z}_2^n$. The code is said to be linear because the sum of two codewords is another codeword, since the subgroup is closed under addition.

Example 8.1 *Let $n = 4$, and consider the group $G = \mathbb{Z}_2^4$ of binary vectors of length 4 under the operation of addition of binary vectors modulo 2. There are $2^4 = 16$ vectors in the group G. As an illustration of a linear code, we may consider a code C to be the subgroup consisting of the vectors*

$$C = \{(0,0,0,0), (1,0,1,1), (1,1,0,1), (0,1,1,0)\}.$$

We note that the sum (calculated modulo 2) of any two of these codewords gives another codeword in the code C. This is an important concept in algebraic coding theory, and so we provide the full addition table to illustrate this point. In the table, we have omitted the parentheses and commas when listing the various vectors.

+	0000	1011	1101	0110
0000	0000	1011	1101	0110
1011	1011	0000	0110	1101
1101	1101	0110	0000	1011
0110	0110	1101	1011	0000

For a binary vector $\mathbf{v} = (a_1, \ldots, a_n)$, the **weight** $wt(\mathbf{v})$ of $\mathbf{v}$ is the number of non-zero coordinates in $\mathbf{v}$.

Given two vectors $\mathbf{u}_1 = (a_1, \ldots, a_n)$ and $\mathbf{u}_2 = (b_1, \ldots, b_n)$, the **distance** $d(\mathbf{u}_1, \mathbf{u}_2)$ between $\mathbf{u}_1$ and $\mathbf{u}_2$, is defined to be the number of coordinates where $\mathbf{u}_1$ and $\mathbf{u}_2$ differ.

Example 8.2 *With $n = 4$, consider $\mathbf{u}_1 = (0,1,1,0)$ and $\mathbf{u}_2 = (1,1,1,0)$. Then $wt(\mathbf{u}_1) = 2$, $wt(\mathbf{u}_2) = 3$, and $d(\mathbf{u}_1, \mathbf{u}_2) = 1$.*

Similarly, the weights of the four codewords from Example 8.1 are 0, 3, 3, and 2, respectively.

The reader should check that, in a linear code, the distance between two vectors $\mathbf{u}_1$ and $\mathbf{u}_2$ is the same as the weight of their difference, i.e.,

$$d(\mathbf{u}_1, \mathbf{u}_2) = wt(\mathbf{u}_1 - \mathbf{u}_2);$$

see Exercise 8.6.10.

It turns out that the distance between vectors is an extremely important one in algebraic coding theory, as it will tell us how many errors a given code can correct or detect; see Theorem 8.3.

The **minimum distance** d_C of a linear code C is defined to be

$$d_C = \min_{\mathbf{c}_1, \mathbf{c}_2 \in C, \mathbf{c}_1 \neq \mathbf{c}_2} d(\mathbf{c}_1, \mathbf{c}_2).$$

As a result of Exercise 8.6.10, the minimum distance can also be found by calculating

$$d_C = \min_{\mathbf{c} \neq \mathbf{0} \in C} wt(\mathbf{c}).$$

Note that our linear code C from Example 8.1 has minimum distance 2.

We now prove an important result in coding theory. The next theorem shows why one wants to build codes with large minimum distances.

Theorem 8.3 *(i) A linear code C can* **correct** *t errors if the minimum distance d_C of the code satisfies $d_C \geq 2t + 1$;*

(ii) A linear code C can **detect** *s errors if the minimum distance $d_C \geq s + 1$.*

Proof: For part (i), assume that $d_C \geq 2t + 1$. Suppose that a codeword $\mathbf{c}$ has been transmitted and the vector $\mathbf{y}$ is received in which t or fewer errors have occurred, so that $d_C(\mathbf{c}, \mathbf{y}) \leq t$. Let $\mathbf{c}'$ be another codeword different from the codeword $\mathbf{c}$. Then $d_C(\mathbf{c}', \mathbf{y}) \geq t + 1$, for if not, $d_C(\mathbf{c}', \mathbf{y}) \leq t$ implies that

$$d_C(\mathbf{c}, \mathbf{c}') \leq d_C(\mathbf{c}, \mathbf{y}) + d_C(\mathbf{y}, \mathbf{c}') \leq t + t = 2t,$$

which contradicts the fact that $d_C \geq 2t + 1$. Thus $\mathbf{c}$ is indeed the codeword closest to $\mathbf{y}$ and our decoding corrects the t errors.

To prove part (ii), assume that $d_C \geq s + 1$. Suppose that the codeword $\mathbf{c}$ is transmitted and that s or fewer errors are introduced during the transmission process. Then the received vector cannot be a different codeword, and so the errors have been detected. ∎

From Example 8.1, we note that the code C with $n = 4$ has $d_C = 2$. Thus, the code can detect one error, but it cannot correct any errors. Hence, the code C from Example 8.1 is not very beneficial. However, we can alter our codes so that they can correct, and detect, more errors. These ideas will be discussed below once we have illustrated how to decode received vectors.

The reader may ask, what is the probability of being able to decode correctly? We now provide a brief discussion, to indicate that this can be very high.

We first assume that errors occur independently with a fixed probability of $1 - p$. The probability $P(i)$ of having i errors in a received vector of length n is given by the formula

$$P(i) = \binom{n}{i}(1 - p)^i p^{n-i}.$$

The following table shows these probabilities when $n = 5$ and $p = 0.9$.

i	$P(i)$
0	0.59049
1	0.32805
2	0.07290
3	0.00810
4	0.00045
5	0.00001

In this case there is less than a 9/1000 chance that a received vector will arrive with 3 or more errors. Thus a code with the ability to correct only two errors will be able to correctly decode almost all received vectors.

If C is a linear code of length n with 2^k codewords and minimum distance d, we say that the code is an $[n, k, d]$ linear code. In a linear $[n, k, d]$ code C, the positive integer k is called the **dimension** of C. We will see later that this really is the dimension of the code when the code is viewed as a vector space over the binary field of two elements.

8.2 Hamming codes

We now construct an infinite class of codes known as **Hamming** codes, named after Richard Hamming (1915–1998), who studied them in 1950 while working at Bell Telephone Laboratories in New Jersey. (Prior to his extensive career at Bell Labs, Hamming was a member of the Manhattan Project at Los Alamos Laboratories during World War II.) His paper [12] provided the start of modern error-correcting coding theory.

In order to begin constructing Hamming codes, we let $m \geq 2$ be an integer. We first form a $(2^m - 1) \times m$ matrix H whose rows consist of all of the non-zero binary vectors of length m. There are, of course, $2^m - 1$ non-zero binary vectors of length m.

For convenience, we list the $m \times m$ identity matrix at the bottom of the matrix H, and then include the remaining non-zero binary vectors of length m listed as rows in H. These remaining $2^m - 1 - m$ rows of H can be listed in any order.

We may thus assume that our matrix H has the form

$$H = \begin{pmatrix} B \\ I_m \end{pmatrix},$$

where B is a

$$(2^m - 1 - m) \times m$$

binary matrix, and I_m denotes the $m \times m$ identity matrix with 1's on the main diagonal and 0's elsewhere.

The matrix H is called a **parity check** matrix for the binary Hamming code H_m. As a small illustration, a parity check matrix for the binary Hamming code H_3 is given by the 7×3 matrix

$$H = \begin{pmatrix} 0 & 1 & 1 \\ 1 & 0 & 1 \\ 1 & 1 & 0 \\ 1 & 1 & 1 \\ 1 & 0 & 0 \\ 0 & 1 & 0 \\ 0 & 0 & 1 \end{pmatrix}.$$

We next form a **generator matrix** G for the Hamming code H_m by first forming a

$$(2^m - 1 - m) \times (2^m - 1)$$

binary matrix

$$G = (I_{n-m} : B) = (I_{2^m - 1 - m} : B).$$

Generator matrices for general codes will be discussed further in the next section; here we just focus on Hamming codes so as to be able to give the reader a concrete set of codes to study.

The matrix G is a generator matrix for a Hamming code H_m whose parameters turn out to be $[2^m - 1, 2^m - 1 - m, 3]$. The length n of the binary Hamming code H_m is easily seen to be $n = 2^m - 1$ since this is the number of non-zero binary vectors of length m. The dimension k of the code is given by $k = 2^m - 1 - m$ since the identity matrix I_m has m linearly independent rows.

The only parameter left to determine is the minimum distance d_{H_m} of the binary Hamming code H_m. It is not too hard to show that the minimum distance is $d_{H_m} = 3$ for any Hamming code H_m. The proof would take us a bit out of our way so we simply refer the interested reader to [14] for a proof of this result. From Theorem 8.3 above, for any $m \geq 2$, the binary Hamming code H_m can thus correct one error or detect two errors.

As an illustration, we construct a generator matrix for the binary Hamming code H_3. To this end we first build a parity check matrix for the Hamming code H_3. Following the discussion above, we can build this matrix as follows:

$$H = \begin{pmatrix} 0 & 1 & 1 \\ 1 & 0 & 1 \\ 1 & 1 & 0 \\ 1 & 1 & 1 \\ 1 & 0 & 0 \\ 0 & 1 & 0 \\ 0 & 0 & 1 \end{pmatrix}.$$

As indicated earlier in the discussion of a generator matrix, a matrix B

for the binary Hamming code H_3 can thus given by

$$B = \begin{pmatrix} 0 & 1 & 1 \\ 1 & 0 & 1 \\ 1 & 1 & 0 \\ 1 & 1 & 1 \end{pmatrix}.$$

Thus the matrix B simply contains the first four rows of the parity check matrix H.

Then a generator matrix for the $[7, 4, 3]$ binary Hamming code H_3 can be given as

$$G = \begin{pmatrix} 1 & 0 & 0 & 0 & 0 & 1 & 1 \\ 0 & 1 & 0 & 0 & 1 & 0 & 1 \\ 0 & 0 & 1 & 0 & 1 & 1 & 0 \\ 0 & 0 & 0 & 1 & 1 & 1 & 1 \end{pmatrix}.$$

The matrix G is simply constructed by placing a copy of the 4×4 identity matrix to the left of the matrix B. Note that this same generator matrix is given in Exercise 8.6.8 where the reader is asked to determine several properties of this code.

We have now constructed a binary linear code which has parameters $[7, 4, 3]$. As will be seen in the next section (which describes how to encode messages in any linear code), we can use the generator matrix G in the above Hamming code H_3 to encode messages and to obtain all of the codewords in the code.

In Section 8.4 we will discuss two methods to decode received vectors using the parity-check matrix for any linear code. In particular, the above 7×3 parity-check matrix H will be very useful for decoding received messages when using the binary linear Hamming code.

Finally in Section 8.5 we will briefly discuss perfect codes. It will be seen that Hamming codes provide the only non-trivial infinite family of perfect codes.

8.3 Encoding

Recall that n is the length of our code, i.e., the number of coordinates in each codeword. If we have 2^k codewords in our code, the positive integer k is called the **dimension** of the code. The word "dimension" is used because a binary linear code C really is a vector space over the binary field $\mathbb{Z}_2$. Thus the code can be viewed as the set of all 2^k binary vectors of length n generated from a basis of the vector space.

Let d be the minimum distance of the code. As mentioned above, we denote a linear code of length n, dimension k, and minimum distance d by saying that the code C is an $[n, k, d]$ linear code. In cases where we don't know the minimum distance of the code C, we will simply use the notation that C is an $[n, k]$ code to indicate that the code C is linear and that it has length n and dimension k.

How does one construct an $[n, k, d]$ linear code? In general it is very difficult to know if an $[n, k, d]$ linear code for a given length n, dimension k, and minimum distance d even exists.

One method that is very helpful to construct linear codes involves the use of a **generator matrix**, an example of which was mentioned above for the binary Hamming code H_m for $m \geq 2$. We now generalize the illustration from the previous section to demonstrate the construction of such codes.

We first fix the length n. Then one needs to have k linearly independent binary vectors of length n that will form a basis for the code. How does one find such linearly independent vectors?

One method is to start with the $k \times k$ identity matrix I_k consisting of all ones on the main diagonal and zeros elsewhere. Then choose a $k \times (n - k)$ binary matrix A. Now form a $k \times n$ binary matrix $G = (I_k : A)$. In theory one could use any $k \times (n - k)$ binary matrix A. However, as will be seen shortly, if one wants the constructed code to be a good one, i.e., to be able to correct numerous errors, the matrix A will have to be constructed in some very special way. We can't expect a randomly chosen matrix to lead to a high-quality code.

The matrix G is a generator matrix because from G, one can generate, or construct, each of the 2^k codewords by taking all of the binary linear combinations of the rows of G. The set of all linear combinations of the rows of G is the **row space** of the matrix G.

For example, consider the generator matrix

$$G = \begin{pmatrix} 1 & 0 & 1 & 1 \\ 0 & 1 & 1 & 0 \end{pmatrix}.$$

Here our code has length 4 and dimension 2. The

$$2^k = 2^2 = 4$$

codewords can be obtained as all binary linear combinations of the rows of G:

$$
\begin{aligned}
0(1, 0, 1, 1) &= (0, 0, 0, 0) \\
1(1, 0, 1, 1) &= (1, 0, 1, 1) \\
1(0, 1, 1, 0) &= (0, 1, 1, 0) \\
1(1, 0, 1, 1) + 1(0, 1, 1, 0) &= (1, 1, 0, 1).
\end{aligned}
$$

To obtain an $[n, k]$ linear code one could also take any set of k linearly independent vectors of length n and use them to form the rows of a generator matrix. The matrix will still have rank k.

The **systematic form** $G = (I_k : A)$ can always be obtained by row reducing any $k \times n$ matrix of rank k. As we will now show, the systematic form for a generator matrix is very convenient for encoding messages.

Using an $[n, k]$ linear code C, we will always have exactly 2^k possible messages that may need to be encoded. How does one encode a message $\mathbf{m} = (a_1, \ldots, a_k)$? We view the message as a $1 \times k$ vector and perform the binary matrix calculation $\mathbf{m}G$ to obtain the corresponding codeword $\mathbf{c}$.

Note that the message vector $\mathbf{m}$ is a $1 \times k$ vector and the generator matrix G is a $k \times n$ matrix, so the product vector $\mathbf{c} = \mathbf{m}G$ (the corresponding codeword) is thus a $1 \times n$ vector.

Using our example with $n = 4$ and $k = 2$, we have $2^2 = 4$ binary messages, which are $(0,0), (1,0), (0,1), (1,1)$. The messages are then encoded as follows:

$$(0,0)G = (0,0) \begin{pmatrix} 1 & 0 & 1 & 1 \\ 0 & 1 & 1 & 0 \end{pmatrix} = (0,0,0,0)$$

$$(1,0)G = (1,0) \begin{pmatrix} 1 & 0 & 1 & 1 \\ 0 & 1 & 1 & 0 \end{pmatrix} = (1,0,1,1)$$

$$(0,1)G = (0,0) \begin{pmatrix} 1 & 0 & 1 & 1 \\ 0 & 1 & 1 & 0 \end{pmatrix} = (0,1,1,0)$$

$$(1,1)G = (0,0) \begin{pmatrix} 1 & 0 & 1 & 1 \\ 0 & 1 & 1 & 0 \end{pmatrix} = (1,1,1,0)$$

The reader should note that, as in the above example, when using a generator matrix in systematic form, the message always occurs as the first k coordinates of the codeword of length n. This happens because of the identity matrix I_k in G.

Matrix products can be calculated very quickly by machine, so this matrix method is a very useful method of encoding. The reader should also check that, whether one uses the matrix multiplication method to encode messages into codewords, or one takes all binary linear combinations of the rows of the generator matrix G, the set of codewords will be the same.

We can thus easily construct codes of length n and dimension k. Just take the identity matrix I_k and an arbitrary $k \times (n - k)$ binary matrix A and form a $k \times n$ generator matrix $G = (I_k : A)$.

Now comes the difficult part of coding theory! How do we form a generator matrix G so as to obtain a code with a large minimum distance? As you will no doubt quickly conclude and as we mentioned above, if we just use any binary matrix A, it seems unlikely, probably very unlikely, that we will end up with a generator matrix G for which the resulting code C has a large minimum distance d_C. Recall from Theorem 8.3 that we would like to have a large minimum distance in order to be able to correct numerous errors.

Given n and k, there is no known method to obtain a generator matrix that will ensure a large minimum distance. Numerous very specialized methods have been used to construct codes with large minimum distance. But in general, given a generator matrix G, it is very difficult to obtain the minimum distance of the resulting code. Researchers have also tried to randomly generate a binary generator matrix and then check the resulting code for its minimum distance. Unfortunately, this process involves significant computing time for large values of n and k.

There is a bound, known as the **Singleton bound**, which implies that in any $[n, k, d]$ linear code C, the minimum distance

$$d_C \leq n - k + 1.$$

Codes that attain this bound with equality are called **minimum distance separable (MDS)** codes. This is an important class of codes; the reader can find a very readable discussion of MDS codes in Chaper 15 of [14].

8.4 Decoding

If one receives a vector, how does one decode the vector to locate the nearest codeword? This is where we make use of Lagrange's Theorem from finite group theory. Recall that Lagrange's Theorem tells us that the number of elements in a subgroup of a finite group must divide the number of elements in the group itself.

Consider our earlier Example 8.1 with $n = 4$. Here the group $\mathbb{Z}_2^4$ contains $2^4 = 16$ vectors. Moreover, that code C contains four codewords. We now form the **coset decomposition** of the group G by C (as it is known by many groups theorists). This same decomposition of G is referred to by coding theorists as the **standard array** for decoding the code C. It may be helpful here for the reader to review the notion of a coset decomposition in Section 2.4 of Chapter 2.

How do we form the standard array for a linear code C?

We list in the first row, in any order, all of the codewords in the code C. We then list in subsequent rows the cosets of C in $\mathbb{Z}_2^n$ where we use the vectors in $\mathbb{Z}_2^n$ of least weight (later called coset leaders) not already listed in the standard array. To save space we will omit the parentheses and the commas in our vectors. Thus the standard array for the code from Example 8.1 can be formed as follows:

$0000 + C$	0000	1011	1101	0110
$0001 + C$	0001	1010	1100	0111
$0010 + C$	0010	1001	1111	0100
$1000 + C$	1000	0011	0101	1110

We note that in this example, we do not list the coset $0100+C$ because this coset is the same as the coset $0010 + C$. Recall from the discussion of cosets in the group theory chapter that two cosets are either disjoint or identical. Thus in our case the coset $0100+C$ would be calculated as the set of elements obtained by adding the element 0100 to the first row of the array to obtain the set of elements

$$0100, 1111, 1001, 0010,$$

which is the same as the elements in the coset $0010 + C$.

When building the standard array, one may ask why we form cosets starting with the second row, by first using the vectors of weight one. And, in general, after trying to use the n binary vectors of weight one, we next use the binary vectors of weight two, continuing if necessary with weight-three vectors, etc., until the standard array is completed.

The reason is that in transmitting vectors, it is more likely that a given received vector has fewer errors than more errors. For example, a received vector is less likely to contain three errors than it is to contain no errors, or one error, or two errors.

These vectors of low weight are called **coset leaders**. As will be noted in the following discussion on how to decode a received vector, the coset leaders will represent the possible errors that have arisen during the transmission process. This method of decoding is often called **nearest neighbor** decoding because we assume that the correct codeword is the one "nearest" or closest to the received vector. The codeword is thus taken to be the codeword that differs from the received vector in the least number of places, i.e., whose distance from the received vector is the smallest.

If we receive a vector $\mathbf{v}$, we search the standard array until we find the vector $\mathbf{v}$ and then we decode to the codeword lying at the top of the column containing the vector $\mathbf{v}$. Recall that in the standard array, each of the binary vectors is listed exactly once.

For example, in the code from Example 8.1, if we receive the vector 1111, we find it in the third row of the standard array. The codeword lying at the top of the column directly above 1111 is 1101, so we assume this is the original message that was sent.

The alert reader will note that the codeword 1011 also differs in one coordinate from the received vector. And the reader would be totally correct!

The real issue in this code is that our code has minimum distance $d = 2$, so from Theorem 8.3 it cannot correct any errors.

From Theorem 8.3, the distance $d = 2$ does, however, allow us to detect one error so that we know that the received vector 1111 is not a codeword. Normally one would still decode to the codeword directly above the received vector. The real reason that we cannot correct this single error is simply the fact that our code only has minimum distance $d = 2$.

We have discussed various properties of several small binary linear codes. We now briefly discuss how to construct a generator matrix for a binary $[n, k]$ linear code C over the binary field F_2. Recall that n is the length of the code and k is the dimension of the code C over the field F_2 so that $1 \leq k \leq n$.

Assume that a generator matrix G has k rows and n columns. Further assume that the matrix G has rank k, i.e., assume that the rows of G are linearly independent over the binary field $\mathbb{Z}_2$ of two elements. Given such a matrix of rank k, we can always row reduce the matrix so it has the form $G = (I_k : A)$. This is called the **systematic** form for the generator matrix G.

We thus have an $[n, k, d]$ linear code where it remains to calculate the minimum distance d of the code. In general, it is hard to find the minimum distance of a code unless the generator matrix or a parity-check matrix is obtained in some very special way.

To build an $[n, k]$ linear code we can, however, always use the identity matrix I_k and then use any $k \times (n - k)$ matrix A to form a generator matrix G. The difficulty with this process is that it is hard to find the minimum distance without calculating the weights of each of the 2^k non-zero binary codewords.

We now provide a larger example to further illustrate the basic coding theory ideas. Consider the generator matrix G with $n = 5$:

$$G = \begin{pmatrix} 1 & 0 & 1 & 1 & 0 \\ 0 & 1 & 0 & 1 & 1 \end{pmatrix}.$$

The code C consists of the row space of G, i.e., the set of all possible binary linear combinations of the rows of G. Thus we have, omitting parentheses and commas,

$$C = \{00000, 10110, 01011, 11101\}.$$

The reader should check that there are no other possible vectors in the code C. This follows from the fact that there are only $2^2 = 4$ linear combinations of the two rows of the matrix G.

Hence the code C has minimum distance $d_C = 3$, and so C can correct one error, since $3 \geq 2(1) + 1$, and it can detect two errors, since $d_C = 3 \geq 2 + 1$.

We now form the standard decoding array for the code C. Note that the first row consists of the codewords in C, and the remaining rows of the standard array consist of the cosets of C.

Note that the first five rows of the array have coset leaders of weight one, while the last two rows of the standard array are produced by using coset leaders of weight two.

$00000 + C$	00000	10110	01011	11101
$00001 + C$	00001	10111	01010	11100
$00010 + C$	00010	10100	01001	11111
$00100 + C$	00100	10010	01111	11001
$01000 + C$	01000	11110	00011	10101
$10000 + C$	10000	00110	11011	01101
$01100 + C$	01100	11010	00111	10001
$11000 + C$	11000	01110	10011	00101

Here again note that we omitted the coset leader 00011 since the coset $00011 + C$ is the same as the coset $01000 + C$ as the reader can check by calculating the vectors in the two cosets and noticing that they are the same. Similarly the coset $00110 + C$ is also omitted since it is the same as the coset $10000 + C$.

The reader can check that each of the $2^5 = 32$ binary vectors of length five is listed exactly once in the above standard array for our code C.

Assume now that the vector 11111 is received. What was the original codeword that was submitted? We search the standard array and see that the vector 11111 lies in the last column of the third row. We thus decode to the codeword in the first row lying directly above the vector 11111; i.e., we decode to the codeword 11101.

Note that this codeword differs in one coordinate (the fourth) from the received vector. Thus the error that was introduced during the transmission process occurred in the fourth coordinate; i.e., the error vector is precisely the coset leader 00010!

Thus what is really happening when we use the standard array to decode received vectors is that the received vector $\mathbf{v}$ satisfies

$$\mathbf{v} = \mathbf{c} + \mathbf{e},$$

where $\mathbf{c}$ is the original codeword and $\mathbf{e}$ is the error that has arisen during the transmission process. The vector $\mathbf{e}$ is the coset leader corresponding to the vector $\mathbf{v}$. Thus by decoding a received vector to be the codeword at the top of that column of the standard array, we are simply saying that, with binary arithmetic,

$$\mathbf{c} = \mathbf{v} - \mathbf{e}.$$

Before closing this section on decoding, we briefly discuss another method of decoding which is often used; it is called **syndrome** decoding.

For an $[n, k]$ linear code, assume that we have a binary $k \times n$ generator matrix $G = (I_k : A)$. We can then build an $n \times (n - k)$ **parity-check** matrix

$$H = \begin{pmatrix} A \\ I_{n-k} \end{pmatrix}.$$

Given a vector $\mathbf{v}$ in $\mathbb{Z}_2^n$, the **syndrome** $S(\mathbf{v})$ of the vector $\mathbf{v}$ is defined to be the matrix product $\mathbf{v}H$, which is a binary vector of length $n - k$. For example, in the binary Hamming code H_3 defined earlier, the syndrome of the vector

$$\mathbf{v} = (1, 0, 0, 0, 0, 0, 0)$$

is the binary vector

$$\mathbf{v}H = (1, 1, 1)$$

of length three.

In fact, for the binary Hamming code H_3 considered earlier, the entire list of syndromes is seen to be the following:

Coset leader	Syndrome
0000000	000
1000000	111
0100000	110
0010000	101
0001000	011
0000100	100
0000010	010
0000001	001

We note in this list of coset leaders and syndromes for the binary Hamming code H_3 that the coset leaders of weight zero and weight one yield the total number of syndromes; namely

$$2^{n-k} = 2^{7-4} = 2^3 = 8.$$

This is exactly the property needed in order for a Hamming code to be **perfect**; see Section 8.5 for a discussion of perfect codes. Recall that in any Hamming code H_m the minimum distance $d_{H_m} = 3$, so the code can correct one error.

Notice that in our earlier example the non-zero coset leaders in the code are precisely the vectors of length $n = 2^m - 1 = 7$, which have weight one.

In general, syndromes have the following properties that make them extremely effective for decoding linear codes.

Theorem 8.4 *(i) A vector* $\mathbf{w}$ *is a codeword of a code* C *if and only if the syndrome* $S(\mathbf{w})$ *is the zero vector of length* $n - k$.

(ii) Two vectors are in the same row of the standard array of the code C *if and only if they have the same syndromes.*

Proof: To prove part (i), note that $\mathbf{w}$ is a codeword if and only if $\mathbf{w}$ has the form $\mathbf{uv}$ with $\mathbf{v} = \mathbf{u}A$ where the generator matrix $G = (I_k : A)$. Hence we have

$$\mathbf{0} = \mathbf{u}A - \mathbf{v}I_{n-k} = \mathbf{u}A + \mathbf{v}I_{n-k} = (\mathbf{uv})H = \mathbf{w}H.$$

Hence $\mathbf{w}$ is a codeword if and only if the syndrome

$$S(\mathbf{w}) = \mathbf{w}H = \mathbf{0}.$$

For part (ii), note that two vectors $\mathbf{u}$ and $\mathbf{v}$ are in the same row of the standard array if and only if they differ by a codeword $\mathbf{w}$, where

$$\mathbf{u} = \mathbf{v} + \mathbf{w},$$

i.e., if and only if

$$\mathbf{u} - \mathbf{v} = \mathbf{w} \in C.$$

Since

$$(\mathbf{u} - \mathbf{v})H = \mathbf{u}H - \mathbf{v}H$$

and $\mathbf{w}H = \mathbf{0}$ with $\mathbf{w} \in C$, we have that $\mathbf{u}$ and $\mathbf{v}$ are in the same row if and only if $\mathbf{u}H = \mathbf{v}H$, which completes the proof. ∎

Note that part (ii) implies that two vectors are in different rows of the standard array if and only if they have different syndromes. We also note that the syndrome properties described in Theorem 8.4 indeed hold for the syndromes of the Hamming code H_3 calculated above.

In order to decode a received vector $\mathbf{v}$ using syndrome decoding, one first calculates the syndrome $S(\mathbf{v})$ and then searches the list of all syndromes until we find the syndrome $S(\mathbf{v})$. We then find the corresponding coset leader $\mathbf{a}$ and decode the received vector $\mathbf{v}$ to be the codeword $\mathbf{c}$ where $\mathbf{v} - \mathbf{a} = \mathbf{c}$. We are thus assuming that the error that occurred in the transmission process is the coset leader (recall that fewer errors are more likely than more errors).

In general, how does one obtain the list of syndromes for a given code? For an $[n, k]$ linear code, in order to determine all of the distinct syndromes, one needs to calculate a list of the 2^{n-k} distinct syndromes, one corresponding to each row of the standard array. We would, however, like to calculate the complete set of syndromes without first building the standard array.

For a large code, this is a not an easy process. For example, for a binary $[100, 60]$ linear code, there are $2^{100-60} = 2^{40}$ distinct syndromes. How are we to find them?

One could proceed as follows. First, consider each binary vector of length n of weight 1 and calculate the syndrome for each of these vectors. Next, check which of these syndromes are distinct, keeping those that are different and discarding any duplicate syndromes.

One then continues with vectors of length n and weight 2, calculating their syndromes and, again, discarding any duplicates in our list of syndromes arising from vectors of weights 1 or 2.

One then continues this process with vectors of weights 3, 4, etc. The difficulty is that as the list of distinct syndromes grows longer, it becomes increasingly more difficult to locate new syndromes.

The use of syndromes avoids having to construct and store the standard array of a code, which can be extremely large. The downside of the syndrome method of decoding is that in a large code, as explained above, it is not easy to construct all of the distinct syndromes.

We now provide an illustration of how to calculate the set of syndromes for a linear code. We note from our previous example where the generator matrix for a binary linear code of length 5, that when using the standard array, the coset of the vector 00011 must be discarded as this coset will be the same as the coset of the vector 01000, i.e., we keep only one copy of each coset. We now show that the same sort of situation arises when calculating syndromes.

To illustrate this, we have the generator matrix

$$G = \begin{pmatrix} 1 & 0 & 1 & 1 & 0 \\ 0 & 1 & 0 & 1 & 1 \end{pmatrix}.$$

Hence the parity-check matrix for the code can be obtained as

$$H = \begin{pmatrix} 1 & 1 & 0 \\ 0 & 1 & 1 \\ 1 & 0 & 0 \\ 0 & 1 & 0 \\ 0 & 0 & 1 \end{pmatrix}.$$

In general, if a generator matrix $G = (I_k : A)$ for some binary matrix A, then in a binary linear code the parity-check matrix H for the same code is given by

$$H = \begin{pmatrix} A \\ I_{n-k} \end{pmatrix}.$$

See the discussion in the Hamming code material that illustrates this connection between generator matrices and parity-check matrices.

Returning to our example, the syndrome

$$S(01000) = (01000)^T H = (011),$$

where the T denotes the transpose of the vector (01000).

Similarly, the syndrome

$$S(00011) = (00011)^T H = (011),$$

which is the same syndrome that we just calculated. Hence this extra copy of the syndrome (011) must be discarded. In our list of syndromes we want to have each of the $2^{n-k} = 2^{5-2} = 2^3 = 8$ distinct syndromes listed exactly once.

We refer the reader to [14] and [29] for further discussion of syndrome decoding.

8.5   Further study

Due to space concerns in this chapter, we have left out many very interesting and important coding theory topics. These topics include bounds on the $[n, k, d]$ parameters for the existence of linear codes. For example, given a length n, dimension k, and distance d, is there a binary linear code with parameters $[n, k, d]$? This is a fascinating but also a very difficult unsolved question.

Other missing topics in this chapter include discussion of the probabilities of correct decoding using linear $[n, k]$ codes, codes constructed over general finite fields (in particular, Hamming codes constructed over a finite field), cyclic codes, equivalent codes, and perfect codes.

Perfect codes are very rare, but the Hamming codes provide an infinite class of such codes. We now give a very brief discussion of perfect codes.

There are 2^n binary vectors of length n. An $[n, k, 2t + 1]$ linear t error-correcting code must satisfy the Hamming bound

$$2^k \left[1 + \binom{n}{1} + \binom{n}{2} + \cdots + \binom{n}{t} \right] \leq 2^n.$$

This follows from the fact that the code has 2^k codewords, and the fact that a sphere of radius t must contain exactly the number of vectors inside the brackets on the left-hand side. Then the product of these two terms must be at most 2^n since there are only 2^n binary vectors of length n.

A code is **perfect** if the inequality above is actually an equality, i.e., the product on the left is equal to 2^n.

The reader should show that the $[7, 4, 3]$ binary Hamming code H_3 constructed in Section 8.2 is perfect.

It turns out that all Hamming codes (even those constructed over non-binary fields) are perfect. We will now show that all binary Hamming codes are perfect.

Recall from Section 8.2 that for an integer $m \geq 2$, the binary Hamming code H_m has parameters

$$[n, k, d] = [2^m - 1, 2^m - 1 - m, 3].$$

Hence we know from the Hamming bound above that

$$2^k \left[1 + \binom{n}{1} \right] = 2^{2^m - 1 - m}[1 + 2^m - 1]$$

which simplifies to

$$2^{2^m - 1 - m} 2^m = 2^{2^m - 1} = 2^n.$$

so the binary Hamming code is indeed a perfect code.

It is known that any non-trivial perfect code has the parameters of a Hamming or a Golay code; see page 102 of [14]. We have already discussed Hamming codes. There are only two linear Golay codes with parameters $[23, 12, 7]$ over the binary field and a $[11, 6, 5]$ linear code over the field of three elements.

We will not go into a detailed discussion of these codes except to give a generator matrix $G = (I_6 | A)$ for the $[11, 6, 5]$ linear Golay code over the field $\mathbb{Z}_3$, where the matrix A can be given as

$$A = \begin{pmatrix} 0 & 1 & 1 & 1 & 1 \\ 1 & 0 & 1 & 2 & 2 \\ 1 & 1 & 0 & 1 & 2 \\ 1 & 2 & 1 & 0 & 1 \\ 1 & 2 & 2 & 1 & 0 \\ 1 & 1 & 2 & 2 & 1 \end{pmatrix}.$$

The matrix I_6 is the usual 6×6 identity matrix with ones on the main diagonal and zeros elsewhere.

A generator matrix for the $[23, 12, 7]$ binary linear Golay code can be given as $(I_{12}|B)$, where the matrix B is a 12×11 binary matrix which, for example, can be found on page 100 of the book [25].

The Hamming codes thus provide the only infinite class of perfect codes; see [14].

We refer to the textbooks [14], [15], [23], and [29] for a discussion of these and many other coding theory topics.

8.6 Exercises

In other chapters of this text we have provided a set of exercises within each section of the given chapter. However, in this chapter, because the coding theory ideas from the various sections are so interrelated, we have simply provided the full set of coding theory exercises here.

1. Determine the length, dimension, and minimum distance for the code with generator matrix

$$G = \begin{pmatrix} 1 & 0 & 0 & 1 \\ 0 & 1 & 1 & 0 \end{pmatrix}.$$

   How many codewords are in this code? List each of the codewords.

2. Let C be the binary linear code having the generator matrix

$$G = \begin{pmatrix} 1 & 0 & 1 & 1 & 1 \\ 0 & 1 & 0 & 1 & 1 \end{pmatrix}.$$

   List all of the codewords of the code C and find the minimum distance of the code. How many errors can this code correct? How many errors can it detect?

3. Determine the list of all coset leaders and syndromes for the binary code C in Exercise 8.6.1.

4. Determine the list of all coset leaders and syndromes for the code in Exercise 8.6.2.

5. Determine the standard array, list of all coset leaders, and syndromes for the binary linear code C obtained from the binary generator matrix

$$G = \begin{pmatrix} 1 & 0 & 0 & 1 & 1 & 0 & 1 \\ 0 & 1 & 0 & 1 & 0 & 1 & 1 \\ 0 & 0 & 1 & 0 & 1 & 1 & 1 \end{pmatrix}.$$

6. Given a binary vector $\mathbf{v}$ of length n, the **sphere** $S_r(\mathbf{v})$ **of radius** r about $\mathbf{v}$ is the set of all binary vectors of length n that differ from $\mathbf{v}$ in at most r coordinates. Thus

$$S_r(\mathbf{v}) = \{\mathbf{w} \in \mathbb{Z}_2^n | d(\mathbf{w}, \mathbf{v}) \le r\}.$$

Show that the number of binary vectors in a radius r sphere $S_r(\mathbf{v})$ about a vector $\mathbf{v}$ is

$$1 + \binom{n}{1} + \binom{n}{2} + \cdots + \binom{n}{r}.$$

Recall that the binomial coefficient $\binom{n}{k}$ is defined to be the value $\frac{n!}{k!(n-k)!}$.

7. List all of the vectors in a sphere of radius 2 about the binary vector 11011.

8. Consider the binary generator matrix given by

$$G = \begin{pmatrix} 1 & 0 & 0 & 0 & 0 & 1 & 1 \\ 0 & 1 & 0 & 0 & 1 & 0 & 1 \\ 0 & 0 & 1 & 0 & 1 & 1 & 0 \\ 0 & 0 & 0 & 1 & 1 & 1 & 1 \end{pmatrix}.$$

Determine the number of codewords in the corresponding code C.

List each of the codewords. Determine the $[n, k, d]$ parameters for this code. How many errors can this code correct? (This code is an example of a Hamming code.)

9. Give an example of a generator matrix G for a code C whose minimum weights of the rows of G is larger than the minimum weight of some codeword of G. This exercise shows that to determine the minimum distance of a code C, it is not enough to simply calculate the minimum weight of each row of a generator matrix of G.

10. Show in a linear code C that for any two vectors $\mathbf{x}$ and $\mathbf{y}$ in C,

$$d_C(\mathbf{x}, \mathbf{y}) = wt(\mathbf{x} - \mathbf{y}).$$

11. List the $2^3 = 8$ possible messages for the code whose generator matrix is given in Exercise 8.6.5.

    Determine each of the corresponding 8 codewords by using the matrix multiplication method of encoding, as well as the row space method of building all binary linear combinations of the rows of G.

12. Construct the parity check and generator matrices for the binary Hamming codes H_2 and H_4. Determine the $[n, k, d]$ parameters for these codes. How many errors can these codes correct? How many errors can these codes detect?

13. Two linear $[n, k]$ codes C_1 and C_2 are said to be **equivalent** if the codewords of C_1 can be obtained from the codewords of C_2 by applying a fixed permutation to the coordinate places of all codewords in C_2.

    Assume that G is a generator matrix for a linear code C. Show that any permutation of the rows of G or any permutation of the columns of G gives a generator matrix of a linear code that is equivalent to C.

14. Using the definition of equivalent codes given in the previous exercise, show that the binary linear codes with generator matrices

$$G_1 = \begin{pmatrix} 1 & 1 & 1 & 0 \\ 0 & 1 & 1 & 0 \\ 0 & 0 & 1 & 1 \end{pmatrix}$$

and

$$G_2 = \begin{pmatrix} 1 & 0 & 1 & 1 \\ 0 & 1 & 1 & 1 \\ 1 & 0 & 0 & 1 \end{pmatrix}$$

are equivalent.

Chapter 9

Appendix

In this Appendix we provide some background information on various topics that will be of use in our text. Much of this material will already be known to some students, so we will only provide a brief, sometimes almost terse, presentation. The intent is for the reader to be able to quickly recall some topics for review, not to provide a detailed study of these topics.

9.1 Mathematical induction

The principle of mathematical induction provides a very powerful method which can sometimes be used to prove that a certain property holds for all positive integers from some starting point on. For example, we show below, for all positive integers $n \geq 1$, that

$$1 + 2 + \cdots + n = n(n+1)/2;$$

i.e., the sum of the first n positive integers is $n(n+1)/2$. We first state the principle of mathematical induction.

Theorem 9.1 *(***Mathematical Induction***) Let $P(n)$ be a mathematical statement about the positive integer n. Assume that $P(1)$ holds and whenever $P(k)$ holds so does $P(k+1)$. Then the statement $P(n)$ holds for all positive integers n.*

The base case in Theorem 9.1 is the case $n = 1$, but we can start induction from any positive integer n_0, and if the induction step works, i.e., if the $n = k$

case implies the $n = k + 1$ case, then the property will hold for all positive integers $n \geq n_0$.

Example 9.2 *We now prove the above statement regarding the sum of the first n positive integers. Let $P(n)$ be the statement that for each positive integer n, the sum*

$$1 + 2 + \cdots + n = n(n+1)/2.$$

For $n = 1$ the sum contains just one term, namely 1, which of course equals $1(1+1)/2$, so the base case holds.

We now assume that the statement $P(k)$ holds for some positive integer k; i.e., that

$$1 + 2 + \cdots + k = k(k+1)/2.$$

We next try to prove that the statement $P(k+1)$ holds. To this end we note that

$$
\begin{aligned}
&1 + 2 + \cdots + k + (k+1) \\
=\ & (1 + 2 + \cdots + k) + (k+1) \\
=\ & k(k+1)/2 + (k+1) \quad \text{by the induction hypothesis} \\
=\ & (k+1)(k/2 + 1) \\
=\ & (k+1)(k+2)/2
\end{aligned}
$$

and this is the statement $P(k+1)$. Thus our statement holds for each positive integer $n \geq 1$.

For example, the sum of the first 100 positive integers is seen to be

$$100(101)/2 = 50(101) = 5050.$$

Example 9.3 *As another example to illustrate the principle of mathematical induction, we now prove that for each positive integer n, the expression $2^{2n} - 1$ is divisible by 3.*

For $n = 1$, it is easy to see that $2^2 - 1 = 3$, so the desired property holds for $n = 1$. For the induction hypothesis we assume that $2^{2k} - 1 = 3M$ for some positive integer M. We now notice that

$$
\begin{aligned}
2^{2(k+1)} - 1 &= 2^{2k+2} - 1 \\
&= 4 \cdot 2^{2k} - 1 \\
&= 4 \cdot 2^{2k} - 4 + 3 \\
&= 4(2^{2k} - 1) + 3 \\
&= 4(3M) + 3 \\
&= 3(4M + 1).
\end{aligned}
$$

Thus since 3 divides $2^{2k} - 1$, then 3 divides $2^{2(k+1)} - 1$, so the proof is complete.

For another illustration of a proof using mathematical induction, we refer the reader to the next section, where it is proved that if A is a set containing exactly n elements, then the power set of A, i.e., the set of all subsets of A, contains 2^n subsets.

We now briefly describe a second form of mathematical induction, often called "strong induction." Again $P(n)$ will denote a mathematical statement about the positive integer n. We first check that $P(1)$ holds (or that $P(n_0)$ holds for some positive integer n_0) to get the induction started. We then assume that $P(m)$ holds for all $m \leq k$ and show that $P(k+1)$ holds.

Note that in the earlier form of induction, to prove that $P(k+1)$ is true, we only used the fact that the previous case $P(k)$ was true. In the strong form of induction, we assume that the statement holds for $m = k$, and that it also holds for each value $m \leq k$; thus the reason for the name strong induction. In fact, it can be shown that the two forms of mathematical induction are equivalent, i.e., each implies the other.

Example 9.4 *We now give an example of the use of strong induction. We prove the existence part of the Fundamental Theorem of Arithmetic (see Chapter 1 for a detailed statement of this theorem). Let S be the set of positive integers greater than 1 that are either primes or can be written as products of primes. Clearly $n = 2$ is in S since 2 is a prime. (Note that in this case we are not starting the induction from the case $n = 1$.) Now assume that for some positive integer n, S contains all integers k with $2 \leq k \leq n$. We must show that $n + 1 \in S$.*

If $n + 1$ is a prime, then it is clearly in S because of the definition of the set S. Assume $n + 1$ is not a prime, so it can be written in the form $n + 1 = ab$ where $1 < a < n + 1$ and $1 < b < n + 1$. From the induction hypothesis and the fact that $a < n + 1$ and $b < n + 1$, we know that a and b are in S. Thus each of the values a and b is either a prime or is a product of primes. Thus when we multiply a and b together, the resulting product $ab = n + 1$ will also be a product of primes and the proof is complete.

9.1 Exercises

1. Define a sequence of positive integers a_n by letting $a_1 = 1$ and then letting $a_{n+1} = 2a_n + 1$ for each $n \geq 2$. Show for each positive integer n that $a_n + 1$ is a power of 2.

2. For each positive integer n, show that $n^5 - n$ is divisible by 5.

3. For each positive integer n, show that $3^{2n} - 1$ is divisible by 8.

4. Find a closed formula for the sum of the first n odd positive integers.

5. Is it true, for each positive integer $n \geq 2$, that $2^n - 1$ is a prime number?

6. Show, for every integer $n \geq 0$, that

$$2^0 + 2^1 + 2^2 + \cdots + 2^n = 2^{n+1} - 1.$$

7. Show that

$$1 \cdot 2 + 2 \cdot 3 + 3 \cdot 4 + \cdots + n(n+1) = \frac{n(n+1)(n+2)}{3}$$

for each positive integer n.

8. Define a sequence of numbers by setting $y_0 = 2$ and $y_1 = 5$ along with the recursion

$$y_{n+2} = 5y_{n+1} - 3y_n$$

for each $n \geq 0$. Show, for each integer $n \geq 0$, that

$$2^n y_n = (5 + \sqrt{13})^n + (5 - \sqrt{13})^n.$$

This problem requires a fair bit of algebra!

9.2   Well-ordering Principle

The following principle may appear to be rather obvious, but it is surprisingly useful. We begin by stating this principle.

Theorem 9.5 *(Well-ordering Principle) Let Y be a non-empty set of positive integers. Then Y has a smallest element.*

How could anything that sounds so simple be so useful? We illustrate with an example by proving the Division Algorithm studied in Chapter 1 (Theorem 1.2).

Theorem 9.6 *(Division Algorithm) Let a and b be natural numbers with $a > 0$. Then there are natural numbers q and r with $0 \leq r < a$ so that $b = aq + r$.*

Proof: If $a > b$, we can simply take $q = 0$ and $r = b$. Thus we may assume that $a \leq b$.

If a divides b, then clearly we may take $r = 0$. Hence we assume that $a \leq b$ and that a does not divide b.

Let

$$D = \{b - ak \mid b - ak \geq 0 \text{ with } k \text{ a non-negative number}\}.$$

The set D certainly consists of integers and it is non-empty because $b = b - a \cdot 0$ is in the set D.

Hence by the Well-ordering Principle, the set D contains a smallest element, say r. Since $r \in D$, r must be of the form $r = b - aq$ for some natural number q. We claim that $r < a$, for if not, then $r - a \geq 0$. Hence we would have

$$r - a = (b - aq) - a = b - a(q+1)$$

and thus $r - a$ would be an element of the set D. But this would contradict the fact that r was the smallest element in the set D. Hence we must have $0 \leq r < a$, and the proof is complete. ∎

We note that this argument proves the existence of the elements q and r so that when we do ordinary long division of b by a, we obtain a quotient q and a remainder r with $0 \leq r < a$.

We now use the Well-ordering Principle to prove the following result concerning **composite numbers**, i.e., numbers that are not primes:

Theorem 9.7 *Every composite number is divisible by some prime number.*

Proof: Let $a \geq 2$ be a positive integer. Let S be the set of positive integers ≥ 2 that divide a. The set S is non-empty since a itself is in S. By the Well-ordering Principle, the set S has a least positive integer, say c. If c is not a prime, then c has a smaller divisor ≥ 2, contradicting the fact that c is the smallest element in the set S. Thus c must be a prime and the proof is complete. ∎

As a final example of the use of the Well-ordering Principle, we show that the Well-ordering Principle implies the Principle of Mathematical Induction.

Theorem 9.8 *The Principle of Well-ordering implies the Principle of Mathematical Induction.*

Proof: Let $P(n)$ be a mathematical statement about the positive integer n. Assume that

(i) $P(1)$ is true and

(ii) for any $n \geq 1$, if $P(n)$ is true then $P(n+1)$ is true.

Let S be the set of positive integers $n \geq 1$ for which $P(n)$ is false. The set S must be either be empty or non-empty. If the set S is empty, we are done.

If the set S is non-empty, we use the Well-ordering Principle to obtain a least element $s \in S$. Since $P(1)$ holds, we know that 1 is not in the set S. Hence $s \geq 2$ so that $s - 1$ is positive.

The definition of s as the least element in S implies that $P(s - 1)$ must hold. But we notice that $s = (s-1)+1$, so by our assumption on the statement $P(s)$, we must have that $P(s)$ holds. This is a contradiction of the fact that s is in the set S. Thus the statement $P(n)$ holds for every positive integer n. ∎

It turns out that the two principles of Well-ordering and Mathematical Induction are indeed equivalent, i.e., that each implies the other. In Exercise 9.2.4 the reader will be asked to prove the converse of the above result.

9.2 Exercises

1. Show that any two positive integers a and b have a least common multiple; i.e., an integer m that is a common multiple of both a and b, and that is less than or equal to any other common multiple of a and b.

2. Show that there is no rational number a/b whose square is 5.

3. Show that there is no rational number whose square is k if k is square-free. A **square-free** number is one that is not divisible by the square of any integer greater than 1; 5, 7, 14, and 22 are examples of square-free numbers.

4. Show that the Principle of Mathematical Induction implies the Well-ordering Principle. Hint: Let S be the set of positive integers that contain no least element. We must show that S is empty. Define a set T to be the set of all positive integers n such that n is not greater than or equal to any element of S. Show by induction that T is the set of all positive integers, and hence the set S is empty, to complete the proof.

9.3   Sets

A **set** is a collection of objects. The objects are called **elements** or **members** of the set. We indicate that an element a is a member of the set A by writing $a \in A$.

The **union** of two sets, denoted by $A \cup B$, is the set $\{x | x \in A \text{ or } x \in B\}$, while the **intersection** of the sets A and B, denoted by $A \cap B$, is the set $\{x | x \in A \text{ and } x \in B\}$. The **empty set** $\emptyset$ is the set that does not contain any elements.

Example 9.9 *If $A = \{1, 2, 3, 4\}$ and $B = \{2, 4, 6\}$ then $A \cup B = \{1, 2, 3, 4, 6\}$ and $A \cap B = \{2, 4\}$.*

Two sets are **equal** if they contain the same elements. A set A is a **subset** of a set B if each element in the set A is also an element in the set B. We denote this by writing $A \subseteq B$. A subset A is a **proper** subset of B if A is a subset of B and there is at least one element $b \in B$ with $b \notin A$. We will denote that the set A is a proper subset of the set B by writing $A \subset B$.

We observe that in order to show two sets are equal, we must show that each is a subset of the other, as illustrated in the following.

Example 9.10 *If A is the set of even integers and*

$$B = \{b | b = 2k + 2 \text{ for some integer } k\},$$

then to show $A = B$, we proceed as follows.

Let $a \in A$. Then $a = 2m$ for some integer m. Hence we also have that $a = 2(m - 1) + 2$ so that $a \in B$. Thus $A \subseteq B$. Similarly if $b \in B$ then $b = 2k + 2$ for some integer k and thus $b = 2(k + 1)$, so b is even and thus $b \in A$. Hence $B \subseteq A$ and thus $A = B$, as desired. $\blacksquare$

Two sets A and B are **disjoint** if they have no elements in common, i.e., if $A \cap B = \emptyset$.

The **power set** of a set A, denoted by $\mathcal{P}(A)$, is the set consisting of all subsets of A. One can show that if a set A has n elements, then the power set $\mathcal{P}(A)$ of A contains 2^n subsets of the set A. We now prove this result by induction on the number n of elements in the set A.

If $n = 1$, assume that the set $A = \{a_1\}$. The subsets of A are the empty set and the set $\{a_1\}$ so the result is true for $n = 1$.

We now assume the result is true for a set that contains $n = k$ distinct elements. We may assume by the induction hypothesis that the distinct subsets of A are $T_1, \ldots, T_{2^k}$.

Now consider a set with $n = k + 1$ elements, which we may assume to be the set $\{a_1, \ldots, a_k, a_{k+1}\}$. We form 2^k new subsets by considering the sets

$$T_1 \cup \{a_{k+1}\}, \ldots, T_{2^k} \cup \{a_{k+1}\}.$$

We first observe that these new sets are all distinct because if

$$T_i \cup \{a_{k+1}\} = T_j \cup \{a_{k+1}\}$$

for some i and j, then removing the element a_{k+1}, we would have that the sets $T_i = T_j$, which is a contradiction.

We next observe that none of the sets of the form $T_i \cup \{a_{k+1}\}$ are equal to any of the sets of the form T_j since the sets T_j do not contain the element a_{k+1}.

Finally, we note that we have

$$2^k + 2^k = 2(2^k) = 2^{k+1}$$

distinct subsets of the set A. This completes the inductive proof which proves that the number of subsets of the set A is at least 2^n.

Every subset of A gives a function $f : \{a_1, \ldots, a_n\} \to \{0, 1\}$ defined by $f(a_i) = 1$ if a_i is in the subset, and $f(a_i) = 0$ if a_i is not in the subset. Thus the total number of subsets is at most the total number of distinct functions, which from Section 9.4 will be seen to be 2^n.

Given two sets A and B, the **Cartesian product** of the sets A and B is the set of ordered pairs defined by

$$A \times B = \{(a, b) | a \in A, b \in B\}.$$

As an illustration, if $A = B = \mathbb{R}$ is the set of real numbers, then $A \times B = \mathbb{R}^2$

is the usual xy-plane so often studied in calculus. On the other hand, if $A = \{1, 2\}$ and $B = \{a, b, c\}$ then

$$A \times B = \{(1, a), (1, b), (1, c), (2, a), (2, b), (2, c)\}.$$

We note in passing that if the set A has m elements and the set B has n elements, then the Cartesian product $A \times B$ has mn elements. This multiplicative property explains why we use the term Cartesian "product."

We now briefly consider an important concept in mathematics; namely that of an equivalence relation. Assume that A is a non-empty set. Define $a \in A$ to be related to $b \in A$ if the elements a, b have some common property; for example, if $a = b$.

We say that such a relation is **reflexive** if every $a \in A$ is related to itself. The relation is **symmetric** if whenever a is related to b, then b is related to a. And we say the relation is **transitive** if whenever a is related to b and b is related to c, then the element a is related to c.

Finally, if the relation satisfies all three properties of being reflexive, symmetric, and transitive, then the relation is said to be an **equivalence relation**.

The set of all elements that are equivalent to a given element a will be called the **equivalence class of** a. We will denote the fact that a is related to b by writing aRb, and we will denote the equivalence class of a by $[a]$. Hence $[a] = \{b \mid bRa\}$. We note that since the relation is symmetric, the equivalence class of a can also be obtained as $[a] = \{b \mid aRb\}$.

We now give some examples of relations and equivalence relations.

Example 9.11   1. *In the set $\mathbb{R}$ of real numbers, define aRb if $a = b$. Since $a = a$ for every $a \in \mathbb{R}$, the relation is reflexive. If $a = b$ then clearly $b = a$, so it is also symmetric. Further, if $a = b$ and $b = c$, then certainly we have that $a = c$, so the relation is also transitive. Thus the relation is an equivalence relation. The reader should check that the equivalence class $[a]$ of a real number a is simply the element a itself.*

2. *In the real numbers, define aRb if $a \leq b$. Then $a \leq a$ so it is reflexive, but the relation is not symmetric, since for example, $2 \leq 3$ but $3 \nleq 2$. The relation is, however, transitive, since if $a \leq b$ and $b \leq c$, then $a \leq c$. Because the relation is not symmetric, it is not an equivalence relation.*

3. *Let n be a positive integer and define aRb if*

$$a \equiv b \pmod{n}$$

where $a, b \in \mathbb{Z}$. The reader should check that this relation is indeed an equivalence relation. Moreover, the equivalence classes are just the congruence classes modulo n discussed in Chapter 1, i.e.,

$$[a] \quad = \quad \{b \in \mathbb{Z} \mid b \equiv a \pmod{n}\}$$

$$= \{b \in \mathbb{Z} \mid b = a + kn \text{ for some integer } k\}.$$

Thus, for example, for $n = 6$, $[2] = \{\ldots, -10, -4, 2, 8, 14, 20, \ldots\}$.

9.3 Exercises

1. If $A = \{a, b\}$ and $B = \{1, 2, c\}$ determine the sets $A \cup B, A \cap B$, and $A \times B$.

2. Find all subsets of the set $A = \{2, 4, a\}$.

3. Let A be the set $A = \{1, 3, 5\}$. Determine the $2^3 = 8$ subsets of A. The set of all of these subsets is the power set $\mathcal{P}(A)$ of the set A.

   List the elements of the Cartesian product set $A \times A$. How many elements are in this Cartesian product set? How many elements are there in the power set $\mathcal{P}(A \times A)$?

4. If a set A contains 10 distinct elements, how many elements are in $A \cup A, A \cap A, A \times A$, and the power set $\mathcal{P}(A)$ of A?

5. If $\mathbb{Z}$ denotes the set of integers and $a, b \in \mathbb{Z}$, define a relation R on $\mathbb{Z}$ by defining aRb if $a - b$ is an even integer. Is this relation R an equivalence relation? If so, what are the equivalence classes?

6. Consider the following sets:

$$A = \{1, 2, 3, 4, 5\} \quad B = \{0, 1, 2, 4\} \quad C = \{1, 3, 5\}$$
$$D = \emptyset \qquad\qquad E = \{2\} \qquad\quad F = \{0, 2, 4\}$$

   List all true statements of the form $X \subseteq Y$ and $X \subset Y$ for these six sets.

7. What is wrong with the following argument? Let R be a symmetric and transitive relation on a set A. Let $a \in A$ and let $b \in A$ so that aRb. Then aRb implies bRa, and aRb and bRa imply aRa. Thus if R is symmetric and transitive, it is also reflexive.

8. Let $A = \{a, b, c, d\}$. Let a relation R be defined by

$$R = \{(a, a), (a, b), (b, b), (b, a), (c, d), (c, c), (d, d), (b, c), (a, c), (b, d)\}.$$

   Is the relation R reflexive, symmetric, transitive? Give a reason for each answer.

9. Prove that $(A \cup B) \times C = (A \times C) \cup (B \times C)$ where $\times$ denotes the Cartesian product of sets.

9.4 Functions

A **function** f from a set A (the **domain**) to a set B (the **co-domain**) is a rule that assigns to each element $a \in A$ a unique element $f(a) \in B$. We will denote a function f from the set A to the set B by writing $f : A \to B$.

The **image** or **range** of a function $f : A \to B$ is the subset $\{f(a) | a \in A\}$ of B containing the images of all of the points of A. As will be shown below, the range of a function is not always the same as the co-domain of the function.

We say that a function $f : A \to B$ is **1-1 (injective)** if f maps distinct elements in A to distinct elements in B. The easiest way to check whether a given function has this property is to check whether

$$f(a) = f(b)$$

implies, or requires that, $a = b$. Some examples of 1-1 functions on the real numbers are given by $f(x) = x^3, f(x) = x^7$, and $f(x) = ax + b$ as long as $a \neq 0$.

The function $f : \mathbb{Z} \to \mathbb{Z}$ defined by $f(x) = x^2$ is not 1-1 since, for example, $f(-2) = 4$ and $f(2) = 4$ but $-2 \neq 2$. The function $f : \mathbb{Z} \to \mathbb{Z}$ defined by $f(x) = 3x$ is seen to be 1-1, or injective, since if $f(x) = f(y)$ then $3x = 3y$ and this implies that $x = y$.

We say that a function $f : A \to B$ is **onto (surjective)** if for every $b \in B$, there is an $a \in A$ with the property that $f(a) = b$. Another way of thinking about this notion of whether a function is onto (surjective) is whether every element of B gets "mapped" to by some element of A. We note that a function is onto or is a surjective function if the range of the function is the same as the co-domain.

For some examples of functions on the reals which are onto, we mention $f(x) = x, f(x) = x^3$, and $f(x) = e^x$ as well as the function $f(x) = \log x$. We note that the function on the integers defined by $f(x) = 3x$ is not onto (not surjective) since, for example, there is no integer x with $f(x) = 3x = 2$. Moreover, $f(x) = x^2$ is not a surjective function on $\mathbb{R}$ because, for example, there is no real number a such that $f(a) = a^2 = -3$.

Finally, we say that a function f is a **bijection** if f is both 1-1 and onto, i.e., if f is both injective and surjective.

We now give some additional examples of functions. Let $A = \{1, 2\}$ and $B = \{a, b\}$ be two sets. One function mapping the set A to the set B is given by

$$f(1) = a, f(2) = a.$$

This function f is neither 1-1 nor onto.

On the other hand, the function defined by

$$f(1) = b, f(2) = a$$

is both 1-1 and onto. The reader should write down all possible functions from the set A to the set B. There are $2^2 = 4$ distinct functions.

Further examples of functions called "permutations" will be discussed in the next section. These special functions have the properties of being both 1-1 (injective) and onto (surjective) and thus are bijections.

Let $f : A \to B$ and $g : B \to C$ be two functions. The **composition** $g \circ f$ of f and g is defined as

$$(g \circ f)(x) = g(f(x))$$

for each x in the set A. Thus we operate from right to left and apply the function f first, then we apply the function g to the result. Note that the function $g \circ f : A \to C$.

As an example, suppose that $f(x) = x^2$ and $g(x) = 2x + 1$ are functions defined on the real numbers. Then

$$(g \circ f)(x) = g(f(x)) = 2(x^2) + 1 = 2x^2 + 1.$$

Analogously the composition $f \circ g$ is given by

$$(f \circ g)(x) = f(g(x)) = (2x + 1)^2 = 4x^2 + 4x + 1.$$

We note that these two compositions are different functions; in general, functions do not commute under the operation of composition.

Two sets A and B have the **same cardinality** if there is a bijection (a 1-1, onto mapping) from the set A to the set B. If the two sets are both finite, then they have the same number of elements.

We note that for two infinite sets, one set can be a subset of the other but they can still have the same cardinality. For example, let $\mathbb{Z}$ be the set of integers and let E be the set of even integers. Define a function $f : \mathbb{Z} \to E$ by defining $f(x) = 2x$. We note that f is a 1-1 function since if $f(x) = f(y)$, then $2x = 2y$ and hence $x = y$. Now let $2t \in E$ be an even integer (so that $t \in \mathbb{Z}$). Then $f(t) = 2t$ so that f maps onto the set E of even integers. Since our function f is both 1-1 and onto (injective and surjective), it is a bijection. Thus the set $\mathbb{Z}$ of integers and the set E of even integers have the same cardinality (even though it is clear that E is a proper subset of $\mathbb{Z}$).

One can also show that the set of positive integers and the set of positive rational numbers have the same cardinality. This is the "Cantor diagonalization process." The details are, however, too complicated to illustrate here, so instead we refer the reader to Chapter 9 of [32].

9.4 Exercises

1. Let $A = \{a, b\}$ and $B = \{1, 2, c\}$. Give an example of a function $f : A \to B$. Give examples of functions that are 1-1. How many 1-1 functions are there mapping the set A to the set B? How many total functions are there mapping the set A to the set B? Is there a function mapping the set A to B that is onto? Why?

2. List all mappings of the set $\{1, 2\}$ to the set $\{a, b\}$; of the set $\{1, 2, 3\}$ to the set $\{a, b\}$; of the set $\{a, b\}$ to the set $\{1, 2, 3\}$.

3. Show that the composition of two 1-1 functions is 1-1.

4. Show that the composition of two onto functions is onto.

5. Give an example of a 1-1 mapping of a set which is not onto.

6. Give an example of an onto mapping of a set that is not 1-1.

7. Let a and b be fixed positive integers. Find a bijection between the set $A = \{as \mid s \in \mathbb{Z}\}$ and $B = \{bt \mid t \in \mathbb{Z}\}$.

8. Let $f : [0, \infty) \to [0, \infty)$ be the function defined by $f(x) = \sqrt{x}$. Is f 1-1? Is f onto?

9. Let $f : [0, \infty) \to \mathbb{R}$ be the function defined by $f(x) = \sqrt{x}$. Is f 1-1? Is f onto?

9.5 Permutations

Assume that X is a non-empty set. A function $f : X \to X$ is a **permutation** if f is both 1-1 and onto; i.e., if f is a bijection. We will usually let X denote a finite set, which we may assume to be the set $X = \{1, 2, \ldots, n\}$.

Let S_n denote the set of all permutations on the set X. As illustrated in the exercises, the set S_n contains exactly $n!$ different permutations. Recall that $n!$ represents the product of the first n positive integers, i.e.,

$$n! = n(n-1)\cdots(2)(1).$$

The set S_n is called the **symmetric group** because the set of permutations does indeed form a group using the operation of composition of functions.

Assume that σ and ρ are two permutations in S_n. As with all functions and hence with all permutations, we will write $\sigma \circ \rho$ to denote the permutation computed as $\sigma(\rho(x))$, where as with functions, we always calculate from right to left. We thus apply ρ to the element x to obtain $\rho(x)$. Then we apply σ to the result to obtain $\sigma(\rho(x))$.

We now briefly discuss cycle notation for permutations in S_n. Given a permutation σ on the set $X = \{1, 2, \ldots, n\}$, we compute the "cycle" of an element x by calculating the elements

$$x, \sigma(x), \sigma^2(x), \sigma^3(x), \ldots,$$

continuing until we find the first power, say r, of the permutation σ so that

$\sigma^r(x) = x$. Since our set X only has a finite number n of distinct elements, such a value of r must exist. We then call the ordered tuple

$$(x, \sigma(x), \sigma^2(x), \ldots, \sigma^{r-1}(x)),$$

the **cycle** of x. Since this cycle contains exactly r distinct elements, we call the cycle of length r an r-**cycle**. Cycles of length one are **fixed points**, while cycles of length two are called **transpositions**.

In cycle notation, we assume that the permutation maps each element to the element to its immediate right, and the last element in the cycle maps back to the first. Thus by the cycle (134) we understand that 1 is mapped to 3, 3 is mapped to 4, and finally 4 is mapped to 1.

Note that the cycles (134), (341), and (413) all look different but as cycles, they are all equal. Each cycle indicates that 1 is mapped to 3, 3 is mapped to 4, and 4 is mapped to 1.

To obtain the cycle representation of a permutation ρ we similarly calculate the cycles of each element in the set X using the permutation ρ.

For example, when $n = 6$, consider the permutation defined by

$$f(1) = 3, f(2) = 2, f(3) = 5, f(4) = 6, f(5) = 4, f(6) = 1.$$

In cycle notation this becomes (13546)(2) or (13546) since fixed points are often omitted.

If a number is missing in a cycle, we simply skip that cycle and move to the next. Within a given cycle, we continue to calculate until we obtain the first element of that cycle, and then we put in a closing parenthesis to indicate that we have completed that cycle.

Two cycles are **disjoint** if each element moved in one cycle is fixed in the other. For example, the cycles (13) and (24) are disjoint since 1 and 3 are moved in the first cycle but are fixed in the second cycle. Similarly 2 and 4 are moved in the second cycle, but they are fixed in the first cycle.

It is true that every permutation can be written as a product of disjoint cycles. For example, the permutation

$$(123)(23)(124)$$

is clearly not in disjoint form since 2 is moved in each of the three cycles.

To obtain the disjoint cycle form for this permutation (or for any permutation) we simply start with the value 1 and write this to begin cycle notation as (1. We calculate from the right, doing one calculation in each of the three cycles. In our case, in the rightmost cycle, 1 moves to 2. Then in the middle cycle, 2 moves to 3, and finally in the leftmost cycle, 3 moves to 1, so the total effect is that 1 stays fixed at 1. We denote this by closing the cycle to obtain (1).

We next start with 2 in the rightmost cycle, which is seen to move to 4. In the middle cycle 4 does not appear, so we move on to the leftmost cycle and see that 4 still does not move. Thus we have the start of the cycle (24.

We now use 4 in the rightmost cycle and see that it moves to 1. The value 1 is then fixed in the middle cycle and in the left cycle 1 moves to 2 so our partially completed cycle (24 now becomes closed and can be written as (24).

Since we have not yet accounted for the value 3, we start with (3 and then working from right to left, 3 stays at 3, in the middle 3 goes to 2, and in the left cycle 2 goes to 3, so the completed cycle is (3), i.e., 3 is another fixed point.

We have now accounted for each of the four values, so the permutation can be written as a product of the disjoint cycles

$$(1)(24)(3).$$

One often deletes the fixed points and then just writes the disjoint form as (24), where it is understood that the values 1 and 3 are both fixed.

Next we illustrate a few calculations using cycle notation, keeping in mind that we always work from right to left and we do one calculation inside each cycle.

Let $\sigma = (2143)$ and $\rho = (32)$ be two permutations in S_4. We calculate the permutation $\sigma\rho$ as follows. Starting from the rightmost cycle, 1 goes to 1 and then in the next cycle to the left, 1 goes to 4, so 1 ends up at 4. Similarly 4 goes to 4 in the right cycle and then 4 goes to 3 in the left cycle, so 4 ends at 3. Finally in the right cycle 3 goes to 2, and in the left cycle 2 goes to 1, so 3 ends at 1.

We have thus completed the cycle (143). We also check that 2 goes to 3 in the right cycle and 3 goes to 2 in the left cycle, so 2 goes to 2 in the product. Thus

$$\sigma\rho = (2143)(32) = (143)(2)$$

in disjoint form. Similarly the reader should check that

$$\rho\sigma = (32)(2143) = (142)(3).$$

Thus $\sigma\rho \neq \rho\sigma$ in this example, so that permutations do not commute under the operation of composition.

Similarly $\rho^2 = (32)(32) = (1)(2)(3)(4)$, which is the identity map on the set $X = \{1, 2, 3, 4\}$.

9.5 Exercises

1. Show that the composition of two permutations on a set X is another permutation on the set X.

2. Show that there are exactly $n!$ different permutations on a set containing n distinct elements.

3. List each of the 6 permutations defined on the set $\{3, 5, 7\}$.

4. If $\sigma = (243), \tau = (142)$, and $\delta = (2143)$ are permutations defined on the set $\{1, 2, 3, 4\}$, compute each of the permutations $\sigma \circ \tau, \tau \circ \sigma, \delta^2 = \delta \circ \delta$, and $\sigma \circ \delta \circ \tau$.

5. For the permutation $\sigma = (423)$, compute the permutation σ^4.

6. For the permutation $\sigma = (423)$, calculate the permutation σ^{-1}, the inverse of the permutation σ. Recall that the inverse σ^{-1} of a permutation σ is defined by

$$\sigma(\sigma^{-1}(x)) = \sigma^{-1}(\sigma(x)) = x$$

for all x in the set X on which the permutations are defined.

7. Show that the inverse of a permutation is itself a permutation.

8. For the permutation $\sigma = (423)$, calculate the permutation σ^{-2}. Recall that this can be calculated in two different ways: as $(\sigma^{-1})^2$ or as $(\sigma^2)^{-1}$.

9.6 Matrices

Here we briefly discuss some basic properties of matrices. For many more details and proofs of some of these results, we refer the reader to [2].

We restrict our attention to square matrices with real coefficients. Assume that A is an $n \times n$ matrix whose entries are all real numbers. We may write A in the form $A = (a_{ij})$ where a_{ij} denotes the element of A that lies in row i and column j for $1 \le i, j \le n$. Assume that $B = (b_{ij})$ is another $n \times n$ matrix.

The matrices A and B are **equal** if $a_{ij} = b_{ij}$ for each $1 \le i, j \le n$; i.e., if the corresponding entries of the two matrices are the same.

We can add two matrices in the rather obvious way by adding corresponding elements so that

$$A + B = (a_{ij}) + (b_{ij}) = (a_{ij} + b_{ij}), 1 \le i, j \le n.$$

One can of course also subtract matrices by replacing the plus signs in the above line by minus signs.

To multiply matrices, one might think that we simply multiply the corresponding entries as we did for addition and subtraction. It turns out that such a process is not very useful, so instead, we multiple matrices using a more complicated procedure. This method will then correspond to the composition of linear transformations; see [2] for details.

Consider the 2×2 matrices

$$A = \begin{pmatrix} 1 & -4 \\ -2 & 3 \end{pmatrix} \quad \text{and} \quad B = \begin{pmatrix} 3 & 1 \\ 1 & -4 \end{pmatrix}.$$

We employ an "across/down" strategy of multiplying and adding as follows. In order to calculate the matrix product AB, we go across the rows of A and down the columns of B multiplying and adding as now illustrated:

$$\begin{pmatrix} (1 \times 3) + (-4 \times 1) & (1 \times 1) + (-4 \times -4) \\ (-2 \times 3) + (3 \times 1) & (-2 \times 1) + (3 \times -4) \end{pmatrix} = \begin{pmatrix} -1 & 17 \\ -3 & -14 \end{pmatrix}.$$

We now illustrate this procedure of matrix multiplication for a pair of 3×3 matrices. We calculate

$$AB = \begin{pmatrix} 1 & -1 & 0 \\ 2 & -2 & 1 \\ 1 & 2 & 2 \end{pmatrix} \begin{pmatrix} 3 & -1 & 2 \\ 0 & 2 & 0 \\ 1 & 3 & -4 \end{pmatrix} =$$

$$\begin{pmatrix} 1 \times 3 + -1 \times 0 + 0 \times 1 & 1 \times -1 + -1 \times 2 + 0 \times 3 & 1 \times 2 + -1 \times 0 + 0 \times -4 \\ 2 \times 3 + -2 \times 0 + 1 \times 1 & 2 \times -1 + -2 \times 2 + 1 \times 3 & 2 \times 2 + -2 \times 0 + 1 \times -4 \\ 1 \times 3 + 2 \times 0 + 2 \times 1 & 1 \times -1 + 2 \times 2 + 2 \times 3 & 1 \times 2 + 2 \times 0 + 2 \times -4 \end{pmatrix}$$

$$= \begin{pmatrix} 3 & -3 & 2 \\ 7 & -3 & 0 \\ 5 & 9 & -6 \end{pmatrix}.$$

Using the same method to calculate the matrix product BA, we note that

$$BA = \begin{pmatrix} 3 & 3 & 3 \\ 4 & -4 & 2 \\ 3 & -15 & -5 \end{pmatrix},$$

so that $AB \neq BA$. Thus under multiplication, matrices do not, in general, commute.

An $n \times n$ matrix A has **rank** n if the n rows of the matrix are linearly independent over the real numbers. A similar result holds for matrices whose entries are elements in any field. The notions of linearly independent and linearly dependent vectors are discussed in Chapter 6.

A matrix A is said to be **non-singular** if the matrix A has rank n. A matrix A is **singular** if it is not non-singular, i.e., A is singular if the rows of A are linearly dependent as vectors. In such a case, the matrix will have rank less than n.

An $n \times n$ matrix is non-singular (**invertible**) if and only if it has rank n. Recall that by the inverse of an $n \times n$ matrix A we mean an $n \times n$ matrix B so that

$$AB = I_n$$

where I_n denotes the $n \times n$, **identity matrix** with ones on the main diagonal and zeros elsewhere throughout the matrix.

We refer to the textbook [2] for many details regarding the theory of matrices.

9.6 Exercises

1. If $A = \begin{pmatrix} 1 & 4 \\ 2 & -3 \end{pmatrix}$ and $B = \begin{pmatrix} -3 & 2 \\ 1 & 4 \end{pmatrix}$ determine $A + B, A - B, AB$, and BA.

2. Are the matrices $A = \begin{pmatrix} 2 & 4 \\ 1 & 3 \end{pmatrix}$ and $B = \begin{pmatrix} -3 & -2 \\ 7 & -1 \end{pmatrix}$ singular or non-singular?

3. If $A = \begin{pmatrix} 3 & -7 \\ 2 & -1 \end{pmatrix}$, calculate A^2 and A^3 where by A^2 we mean the matrix AA.

4. If $A = \begin{pmatrix} 2 & 1 \\ -3 & 4 \end{pmatrix}$ and $B = \begin{pmatrix} 1 & -1 \\ 2 & -1 \end{pmatrix}$, is $AB = BA$?

5. Let $A = \begin{pmatrix} 2 & 1 & -1 \\ -3 & 4 & 1 \\ 2 & -1 & 3 \end{pmatrix}$ and $B = \begin{pmatrix} 1 & -1 & 3 \\ 2 & -1 & 2 \\ 3 & -1 & -2 \end{pmatrix}$. Calculate the matrix products AB, BA, A^2, and B^2.

6. Are the matrices A and B in the previous exercise non-singular?

7. Is the matrix $A = \begin{pmatrix} 1 & 3 \\ 2 & 4 \end{pmatrix}$ singular or non-singular? If A is non-singular, find the inverse matrix A^{-1} of the matrix A, i.e., find a 2×2 matrix $A^{-1} = \begin{pmatrix} a & b \\ c & d \end{pmatrix}$ so that

$$\begin{pmatrix} 1 & 3 \\ 2 & 4 \end{pmatrix} \begin{pmatrix} a & b \\ c & d \end{pmatrix} = \begin{pmatrix} 1 & 0 \\ 0 & 1 \end{pmatrix}.$$

9.7 Complex numbers

We now briefly review a few basic properties involving complex numbers. The **complex numbers** comprise the set $\mathbb{C}$ defined by

$$\mathbb{C} = \{a + bi \,|\, a, b \in \mathbb{R}\}.$$

Here $i = \sqrt{-1}$.

The value a is called the **real part** of $a + bi$ while the value b is called the **imaginary part** of $a + bi$. In addition, we call the complex number $a - bi$ the **complex conjugate** of $a + bi$. The conjugate of $a + bi$ is often denoted by

$$\overline{a + bi} = a - bi.$$

Given two complex numbers, say $a + bi$ and $c + di$, we can add them by calculating

$$(a + bi) + (c + di) = (a + c) + (b + d)i.$$

Subtraction is performed in a similar way by calculating

$$(a + bi) - (c + di) = (a - c) + (b - d)i.$$

To multiply, we use ordinary rules for multiplication along with the fact that $i^2 = -1$ to obtain

$$(a + bi)(c + di) = (ac - bd) + (ad + bc)i.$$

We note that with these definitions, the set of all complex numbers forms a field. In addition, the real numbers are a subfield of the complex numbers; see Chapter 4 for a discussion of fields.

9.7 Exercises

1. Calculate $(2 + 3i) + (6 - 2i)$.

2. Calculate the complex conjugate of $6 - 2i$.

3. Determine the complex conjugate of $(2 - i)^3$.

4. Is the complex conjugate of $(a + bi)^3$ the same as $(a - bi)^3$? If so, prove it.

5. Calculate $(2 + 3i)(6 - 2i)$.

6. Simplify $\frac{(2+4i)^2(3-2i)}{5-i}$ to be in the form $a + bi$.

7. Simplify $(2 - 4i)^3$ to be in the form $a + bi$.

8. Calculate $(2 - i)[(3 + i) + (4 - i)]$ and leave the answer in the form $a + bi$.

9. Calculate the powers i^a for $a = 1, 2, \ldots, 8$ where $i = \sqrt{-1}$.

Chapter 10

Hints and Partial Solutions to Selected Exercises

Here we provide hints and partial solutions for selected exercises. In particular, we provides hints and partial solutions to all odd numbered exercises. In some cases, we give complete details for the solution of a given exercise. In other cases we only provide a hint or idea to get the reader started in the right direction. In some other cases, particularly in some computational exercises, we do not provide too many details, leaving it to the reader to work out most of the various calculations.

We mention that these hints and partial solutions should be used to help the reader work through the various exercises. They should not be simply read to see the solution; rather the reader should look at our hint for a given exercise and then go back with pencil and paper in hand and try to work out the remaining details of the exercise. Don't just copy our solutions; instead use them as helpful study ideas.

Chapter 1

Exercises 1.1

1. • $5 | 635$ since $635 = 5(127)$ and 127 is an integer
   • -5 also divides 635 since $635 = -5(-127)$ and -127 is an integer
   • 48 does not divide 124 since $124 \neq 48k$ for any integer k. One can also see that 48 does not divide 124 by using long division to divide 124 by 48, where the remainder is 28, not 0 as it would be if 48 divided 124.
   • Using long division, we see that $32871 = 96(341) + 135$ so that the remainder is not 0 and hence 341 does not divide 32871.
   • $15m - 10 = 5(3m - 2)$, so yes, 5 divides $15m - 10$
   • $-3m = m(-3)$ so yes, m divides $-3m$
   • No, since $7k + 14m \neq (k + m)x$ where x is an integer
   • $-6k^2 - k = k(-6k - 1)$, so yes, k divides $-6k^2 - k$ since $-6k - 1$ is an integer

3. Since m divides d we have that $d = mk$ for some integer k. Similarly $a = de$ for some integer e since d divides a. Hence we have

$$a = de = (mk)e = m(ke)$$

with ke an integer, so m divides a.

5. By repeated use of the Division Algorithm we have

$$225 = 35(6) + 15$$
$$35 = 15(2) + 5$$
$$15 = 5(3) + 0.$$

Thus the gcd$(35, 225)$ is the last non-zero remainder, which is 5.

7. From the Division Algorithm we have

$$434 = 384(1) + 50$$
$$384 = 50(7) + 34$$
$$50 = 34(1) + 16$$
$$34 = 16(2) + 2$$
$$16 = 2(8) + 0.$$

Hence gcd$(384, 434)$ is the last non-zero remainder, which is 2.

9. We prove this via induction. Let F_k denote the k-th Fibonacci number so that we have the linear recurrence

$$F_{k+2} = F_{k+1} + F_k, k \geq 1$$

with initial values $F_1 = F_2 = 1$. Our aim is to show that for each $k \geq 1$, gcd$(F_k, F_{k+1}) = 1$.

For $k = 1$ we clearly have gcd$(F_1, F_2) = $ gcd$(1, 1) = 1$. By the induction hypothesis we assume that gcd$(F_k, F_{k+1}) = 1$. Assume now that the gcd$(F_{k+1}, F_{k+2}) = d$ for some positive integer d. From the recursion above we also have that $F_{k+2} = F_{k+1} + F_k$. Thus d divides F_{k+2} and d divides F_{k+1}, so that d must also divide their difference, which is F_k. Hence d divides both F_{k+1} and F_k, but our induction hypothesis implies that gcd$(F_k, F_{k+1}) = 1$. Thus $d = 1$ and we are done.

11. We proceed by induction on the positive integer n. For $n = 1$ we clearly have that $4^1 - 1 = 3$ is divisible by 3. Assume that $4^k - 1 = 3M$ for some integer M. Then for $n = k + 1$ we have

$$4^{k+1} - 1 = 4(4^k - 1) + 3 = 4(3M) + 3 = 3(4M + 1) \text{and we are finished.}$$

13. Consider the gcd$(a + 3, a)$ for a positive integer a. Assume this gcd is the positive integer d. Then $d|a$ and $d|(a + 3)$ so that d must divide their difference, which is 3. Since 3 is a prime, we must have that $d = 1$ or $d = 3$. Examples of the former are the pairs $(4, 1), (5, 2), (8, 5), \ldots$. Examples of the latter are the pairs $(6, 3), (9, 6), (12, 9), \ldots$.

Exercises 1.2

1. After using the Sieve of Eratosthenes, your resulting list of primes less than 100 should be

$$2, 3, 5, 7, 11, 13, 17, 19, 23, 29, 31, 37, 41,$$

$$43, 47, 53, 59, 61, 67, 71, 73, 79, 83, 89, 97.$$

3. $384 = 2^7(3)$

5. $6250 = 2(5^5)$

7. Consider the $\gcd(2^5 3^8 5, 3^3 5^5 7^6)$. Since both numbers are already in canonical factored form, we just take each prime to the minimum exponent to which it occurs in the two numbers. Thus the gcd of the two numbers is $2^0 3^3 5^1 7^0 = 3^3 5$.

9. Assume that we are attempting to locate all primes up to the positive integer n. Assume further that n is factored as $n = ab$ with $1 < a, b < n$.

If both $a < \sqrt{n}$ and $b < \sqrt{n}$, then clearly we have that

$$n = ab < \sqrt{n}\sqrt{n} = \sqrt{n^2} = n,$$

a contradiction.

11. Let n be a positive integer. Recall that $n!$ is the product of the first n consecutive positive integers so that $n! = n(n-1)\cdots(2)(1)$.

Now consider the n positive integers

$$(n+1)! + 2, (n+1)! + 3, \ldots, (n+1)! + (n+1).$$

We first note that there are exactly n numbers in the list and that the numbers are indeed consecutive.

Now notice that since 2 divides $n!$, it also divides the number $(n+1)! + 2$ so this first number in the list is even and greater than 2 and thus is not a prime. Similarly, the second number $(n+1)! + 3$ in the list is divisible by 3 so it is not a prime. Continuing, the last number $(n+1)! + (n+1)$ is divisible by $n+1$ so it is also not a prime. Thus none of the listed positive integers is prime and we are done.

Exercises 1.3

1. • $17 - 5 = 12 \neq 9k$ for any integer k so no
   • $33 - 0 = 33$ is divisible by 11 so yes
   • $55 - (-9) = 64$ is divisible by 16 so yes
   • $283 - 177 = 106$ is not divisible by 5 so no
   • $283 - 177 = 106$ is divisible by 2 so yes
   • $220 - 14 = 206$ is not divisible by 6 so no
   • $34 - (-12) = 46$ is divisible by 23 so yes
   • $17 - (-35) = 52$ is not divisible by 9 so no
   • $3m + 1 - (2m + 1) = m$ is divisible by m so yes
   • $3m + 3 - (4m + 3) = -m$ is divisible by m so yes

3. Assume $a - b = kn$ and $c - d = ln$ for integers k and l. Then

$$(a - c) - (b - d) = (a - b) - (c - d) = kn - ln = (k - l)n.$$

5. We need to show that this relation is reflexive, symmetric, and transitive. From properties of congruences we have that $a \equiv a \pmod{n}$ since $a - a = 0 = 0(n)$, so the relation is reflexive.

If $a \equiv b \pmod{n}$, then $a - b = nK$ for some integer K, and hence $b - a = n(-K)$, and hence $b \equiv a \pmod{n}$, so the relation is symmetric.

Finally for transitivity, assume that we have $a \equiv b \pmod{n}$ and $b \equiv c \pmod{n}$. Then we also have that

$$a - c = a - b + b - c = nK + nL = n(K + L)$$

for integers K and L. Thus we have that $a \equiv c \pmod{n}$.

7. Let n be a positive integer. We prove this result by induction on n. For $n = 1$ we have that $8^1 - 1 = 7$ so this initial case holds.

Assume that $8^n - 1 = 7K$ for some integer K. Then we have

$$8^{n+1} - 1 = 8(8^n - 1) + 7 = 8(7K) + 7 = 7(8K + 1),$$

which completes the proof.

9. Assume that $a \equiv b \pmod{n}$ so that $a - b = nK$ for some integer K. Then we have

$$a^e - b^e = (a - b)(a^{e-1} + a^{e-2}b + \cdots + b^{e-1}) = nK(a^{e-1} + a^{e-2}b + \cdots + b^{e-1}),$$

which completes the proof.

Exercises 1.4

1. (a) The $\gcd(5, 12) = 1$, which divides 6, so yes, there is one solution $x = 6$.

(b) $\gcd(5, 20) = 5$ so no solution since $5 \nmid 12$

(c) The $\gcd(4, 20) = 4$ and $4 | 12$ so there are 4 distinct solutions modulo 20, i.e., four distinct solutions x with $0 \le x < 20$. Upon dividing both sides of the congruence by 4, we obtain the congruence $x \equiv 3 \pmod{5}$ whose solution is $x = 3$. The other solutions are then given by $3 + 5 = 8, 3 + 10 = 13$, and $3 + 15 = 18$, where $n/d = 20/4 = 5$.

(d) The $\gcd(12, 16) = 4$, so no solution, since $4 \nmid 14$

(e) The $\gcd(12, 144) = 12$ and $12 | 24$, so yes, there are 12 solutions modulo 144. Dividing the congruence by 12 we obtain the congruence $x \equiv 2 \pmod{12}$, whose solution is $x = 2$. Since $n/d = 144/12 = 12$, the full set of solutions modulo 144 is given by

$$2, 14, 26, 38, 50, 62, 74, 86, 98, 110, 122, 134.$$

3. Using the second congruence we have that $x = 6 + 17k$ for some integer k. Substituting this into the first congruence and reducing the coefficients modulo 11, we obtain $6k \equiv -1 \equiv 10 \pmod{11}$. Thus after a bit of arithmetic, we see that $k = 9$ and so $x = 6 + 17(9) = 159$, which is a solution of both congruences.

5. Using the same ideas as in the solution of Exercise 1.4.3, we assume that $x = 5 + 11k$ for some integer k. Then $5 + 11k \equiv 4 \pmod{9}$ whose solution is $k = 4$, so the solution to the last two congruences becomes $x \equiv 5 + 11(4) = 49$. Then using the first congruence we have $49 + 99k \equiv 5 \pmod{8}$. Simplifying

modulo 8 we see that $k = 4$ and hence our final solution is $x = 49+99(4) = 445$. The reader should check that $x = 445$ satisfies each of the three congruences.

Exercises 1.5

1. (a) Since 5 is a prime and 2 is relatively prime to 5, Fermat's Theorem tells us that $2^4 \equiv 1 \pmod 5$. Thus $2^7 \equiv 2^4 2^3 \equiv 8 \equiv 3 \pmod 5$.

(b) Fermat's Theorem shows that $3^6 \equiv 1 \pmod 7$, so that $3^{14} \equiv (3^6)^2 3^2 \equiv 9 \equiv 2 \pmod 7$.

(c) Since 19 is a prime, Fermat's Theorem (Theorem 1.22) implies that $16^{18} \equiv 1 \pmod{19}$. Hence

$$16^{22} \equiv 16^{18}16^4 \equiv 16^4 \equiv (-3)^4 \equiv 81 \equiv 5 \pmod{19}.$$

3. $10 = 2(5)$ so $\phi(10) = \phi(2(5)) = \phi(2)\phi(5) = (2-1)(5-1) = 4$.
$\phi(24) = \phi(2^3 3) = \phi(2^3)\phi(3) = (2^3 - 2^2)(3-1) = 8$.
Hence $\phi(324) = \phi(2^2 3^4) = \phi(2^2)\phi(3^4) = (2^2 - 2^1)(3^4 - 3^3) = 2(54) = 108$.
$\phi(1024) = \phi(2^{10}) = 2^{10} - 2^9 = 1024 - 512 = 512$.

5. $\gcd(6, 28) = 2$ so no

7. The required elements are those that are relatively prime to 28. There are $\phi(28) = 12$ such elements and they are

$$1, 3, 5, 9, 11, 13, 15, 17, 19, 23, 25, 27.$$

9. Let $n = ak$ where $a > 1$ is odd. Then there is an odd prime p that divides a. Let e be the maximum power of p that divides n and thus n can be written as $n = p^e K$ for some integer K with $\gcd(p, K) = 1$. Hence

$$\phi(n) = \phi(p^e K) = \phi(p^e)\phi(K) = (p^e - p^{e-1})\phi(K),$$

which is even since p is odd and the difference of the two odd integers p^e and p^{e-1} is even.

11. Each element a in the range $1 \le a \le p-1$ is relatively prime to the prime p so each such element a has a multiplicative inverse modulo the prime p.

Exercises 1.6

1. Since $n = pq = 23(29) = 667$, we have that

$$\phi(n) = \phi(23)\phi(29) = 22(28) = 616.$$

The deciphering exponent x satisfies $ax + \phi(n)y = 1$. Using the Euclidean Algorithm (Theorem 1.3) we find that $x = -205$, which we turn into a positive integer by adding 616 to obtain the deciphering exponent $x = 411$.

3. The ciphertext is computed as

$$c \equiv m^a \equiv 6^3 \equiv 216 \pmod{667}.$$

5. The deciphering exponent t satisfies the congruence $st \equiv 1 \pmod{\phi)(n)}$, i.e., $5t \equiv 1 \pmod{36}$. Check that this leads to the solution $t = 29$. We encipher

the message $m = 2$ as $2^s \equiv 2^5 \equiv 32 \pmod{37}$. Since $t = 29$, we decipher the received ciphertext 32 as

$$(32)^{29} \equiv (2^5)^{29} \equiv 2^{145} \pmod{37}.$$

Recall Fermat's Theorem (Theorem 1.22) that since 37 is a prime, $2^{36} \equiv 1 \pmod{37}$. Hence $2^{145} \equiv 2^{144}(2) \equiv 2 \pmod{37}$, so the received value is deciphered as 2.

Chapter 2

Exercises 2.2

1. For a given value of m, we may build the addition table for the group $\mathbb{Z}_m$ of integers modulo m as follows. First label the rows and columns of an $m \times m$ array with the elements $0, 1, \ldots, m - 1$.

For $0 \le x, y \le m - 1$, at the intersection of row x and column y, we place the element $x + y \pmod{m}$. For example, for $m = 2$ we obtain the addition table

+	0	1
0	0	1
1	1	0

3. Assume that e and f are both identity elements in the group G. Then $e = ef = f$.

5. Since $a^2 = e$ for every element $a \in G$, we have that $a = a^{-1}$ for each element $a \in G$. Hence

$$ab = a^{-1}b^{-1} = (ba)^{-1} = ba.$$

Recall that in any group the inverse of a product of two elements is the product of the two inverses in the reverse order.

7. Let σ be the permutation $\sigma = (12)$ so that $\sigma^{-1} = (21)$. Similarly let a permutation τ be defined by $\tau = (123)$ so that $\tau^{-1} = (321)$. Hence we have

$$(\sigma\tau)^{-1} = ((12)(123))^{-1} = ((1)(23))^{-1} = (23)^{-1} = (32).$$

Similarly

$$(12)^{-1}(123)^{-1} = (21)(321) = (13)(2) = (13).$$

9. Since $a \in G$ a group, the element a has an inverse a^{-1} in the group G. Thus if we multiply both sides on the left by the element a^{-1}, we have $b = c$ since $a^{-1}a = e$ is the identity in the group G.

11. For any element a we have $a = eae^{-1}$, so the relation is reflexive.

For symmetry, assume that $b = gag^{-1}$ for some element $g \in G$. Then solving for the element a we see that $a = g^{-1}b(g^{-1})^{-1}$ where we used the fact that $(g^{-1})^{-1} = g$ for any element g in a group G.

For transitivity, assume that $b = gag^{-1}$ and that $c = hbh^{-1}$ for some $g, h \in G$. Then $c = (gh)a(gh)^{-1}$, so c is related to the element a, since $(gh)^{-1} = h^{-1}g^{-1}$.

13. For elements $a, b \in G$, $a * b = a + b + 1$ is an integer, so the operation $*$ is closed.

The operation is associative because of the following two calculations:

$$(a * b) * c = (a + b + 1) * c = (a + b + 1) + c + 1 = a + b + c + 2$$
$$a * (b * c) = a * (b + c + 1) = a + (b + c + 1) + 1 = a + b + c + 2.$$

The reader should check that $e = -1$ is the identity and $a - 2$ is the inverse of the group element a.

15. We prove this result by using induction on n. The result clearly holds when $n = 1$ since $(g^{-1}hg)^1 = g^{-1}hg$.

Assume the result holds for $n = k$ so that we may assume $(g^{-1}hg)^k = g^{-1}h^k g$.

Then we have

$$(g^{-1}hg)^{k+1} = (g^{-1}hg)^k (g^{-1}hg) = g^{-1}h^k g(g^{-1}hg) = g^{-1}h^{k+1}g.$$

17. The operation $\triangle$ is closed since the operations $*$ and $\circ$ are both closed in the groups G and H. This new operation is associative since both of the group operations $*$ and $\circ$ are associative (check this in detail!).

Since G is a group, it has an identity e. Similarly the group H has an identity i. Then the identity for the new group $G\triangle H$ will be the ordered pair (e, i). Check!

Finally if (g, h) is an element (ordered pair) in the new group $G\triangle H$, its inverse will be the ordered pair (g^{-1}, h^{-1}) where g^{-1} is the inverse of the element $g \in G$ and h^{-1} is the inverse of the element $h \in H$. Again, the reader should check this last fact.

Exercises 2.3

1. By Theorem 2.3, a non-empty subset H of a group G is a subgroup if and only if $ab^{-1} \in H$ whenever $a, b \in H$. Let H and K be subgroups of a group G and let $a, b \in H \cap K$ so that $a, b \in H$ and $a, b \in K$. Then since H and K are subgroups, we have that $ab^{-1} \in H$ and it is also in K, so it is in $H \cap K$.

Use induction on the number of subgroups.

3. The element 0 forms a subgroup, as does the entire group $\mathbb{Z}$. The set of all integers of the form $4k$ and $6l$ for integers k and l also form subgroups, as you should check. In fact, as shown in the next exercise, the set of all integers of the form nk can be shown to be a subgroup for any fixed integer k, so we have an infinite number of subgroups of the integers, not just four!

5. Note that $nk - mk = k(n - m)$, so the elements in the given set form a subgroup of the integers.

7. Let $g, h \in N_a$ so that $ga = ag$ and $ha = ah$; then

$$agh^{-1} = gah^{-1} = gh^{-1}a.$$

9. Assume that the group G is a cyclic group generated by some element

$g \in G$. Let $a, b \in G$ so that $a = g^n, b = g^m$ for some integers n and m. Then

$$ab = g^n g^m = g^{n+m} = g^{m+n} = g^m g^n = ba$$

so the group is Abelian.

11. The identity element $e \in Z(G)$ since e commutes with every element in the group. If G is Abelian, then $Z(G) = G$. To show that $Z(G)$ is a subgroup of G, let $a, b \in Z(G)$ so $ax = xa$ and $bx = xb$ for all $x \in G$. Then $x = b^{-1}xb$ and $xb^{-1} = b^{-1}x$, and thus

$$xab^{-1} = axb^{-1} = ab^{-1}x.$$

Exercises 2.4

1. Let H be a subgroup of the group G.

For Part 1, $H = eH$ where e is the identity element.

For Part 2, $a = ae$ so $a \in aH$.

For Part 4, assume that the coset aH is a subgroup of G. Then $e \in aH$ so $e = ah$ for some $h \in H$. Hence $eh^{-1} = a$ so that $h^{-1} = a$ and thus $a \in H$.

3. The elements of H are each of the form $2k$ modulo 16. Thus $2k - 2l = 2(k - l)$ modulo 16, so the set H forms a subgroup.

If a is even, the coset $a + H = H$. If a is odd, the coset

$$a + H = \{1, 3, 5, 7, 9, 11, 13, 15\}.$$

Chapter 3

Exercises 3.1

1. • The set of 3×3 upper triangular matrices forms a non-commutative ring with identity and zero-divisors. The key point to check is whether these matrices are closed under the usual operation of matrix multiplication. This operation holds because of the calculation

$$\begin{pmatrix} a & b & c \\ 0 & d & e \\ 0 & 0 & f \end{pmatrix} \begin{pmatrix} a' & b' & c' \\ 0 & d' & e' \\ 0 & 0 & f' \end{pmatrix} =$$

$$\begin{pmatrix} aa' & ab' + bd' & ac' + bc' + cf' \\ 0 & dd' & de' + ef' \\ 0 & 0 & ff' \end{pmatrix}$$

so the product matrix is also upper triangular.

• These matrices do not form a ring as they are not closed under addition. Note, for example, that the sum of two such matrices would have 2s on the main diagonal.

• The power set with these operations does not form a ring, as there are no additive inverses. It is easy to check that the empty set $\emptyset$ must be the additive identity. But then if A is a non-empty subset of the set X, there is no subset B of X with $A \cup B = \emptyset$.

3. Let $R = \mathbb{Z}_2$ be the ring of integers modulo 2. Then we have

$$(a + b)^2 = a^2 + 2ab + b^2 = a^2 + b^2,$$

since addition is computed modulo 2.

Actually, if one considers the ring $R = \mathbb{Z}_p$ of integers modulo a prime p, a similar result holds where one can simply replace each of the 2's in the above calculation by the prime p.

5. In any ring, we have the following calculation

$$0 = -1(0) = -1(a - a).$$

Thus, $0 = -1(a) + a$ so that

$$0 + (-a) = (-1(a) + a) + (-a) = -1(a) + (a + -a) = -1(a) + 0,$$

or $-a = -1(a)$.

7. Let $R = \mathbb{Z}$ be the ring of integers and let $r = 2$. The element 2 is not invertible since $rk = 2k \neq 1$ for any integer k. Also 2 is not a zero divisor since if $2b = 0$, then $b = 0$ as the ring (integral domain) $\mathbb{Z}$ does not have any zero-divisors.

9. $\phi(28) = \phi(2^2 7) = \phi(2^2)\phi(7) = (2^2 - 2)(7 - 1) = 12$. The invertible elements are those values of a with $1 \leq a \leq 28$ with $\gcd(a, 28) = 1$. The invertible elements are

$$1, 3, 5, 9, 11, 13, 15, 17, 19, 23, 25, 27.$$

11. Let r be an element in R which is not a zero-divisor. Then we have that

$$rs - rt = r(s - t) = 0,$$

so that $s - t = 0$. (Recall that if $s - t \neq 0$, then r would be a zero-divisor.) Hence $s = t$.

13. Note that in a ring, $0 + 0 = 0$ and $0 \cdot 0 = 0$.

15. There are lots of such functions. As examples, we may consider the functions $f_1(x) = 0$, $f_2(x) = x^3 - x$, $f_3(x) = x^9 - x$, $f_4(x) = x^9 - x^3$, $f_5(x) = x^{27} - x^9$.

17. Let $a, b \in R$ so that $a = g + \cdots + g = ng$ for some ring element g and some integer n. Similarly $b = g + \cdots + g = mg$ for some integer m. Thus

$$ab = (g + \cdots + g)(g + \cdots + g) = g^2 + \cdots + g^2.$$

where the sum is over nm terms. But this is clearly the same as summing over mn terms, so we obtain the element ba.

Exercises 3.2

1. It is easy to check that the set of rational numbers is closed under both addition and multiplication, so the rational numbers form a subring of the real numbers.

However, they do not form an ideal because if $\frac{a}{b}$ is rational and r is a real number that is not rational, then $\frac{a}{b}r$ is not a rational number.

3. If $r \in R$, $0(r) = 0 \in \{0\}$ and if $a \in R$, then $ar \in R$, so R is closed under multiplication.

5. We first show that I is a subring of the ring R. To this end, let $x, y \in I$. Now let $a = 0$ and let $b = y$. Then we have $a - b = -b = -y \in I$ since $a - b \in I$ for any $a, b \in I$. Now let $a = x$ and $b = -y$ so that $a - b = x + y \in I$ and I is a subring. The given property that xa and ax are in I for any $x \in I, a \in R$, implies that I is an ideal.

7. Let $a \in I \cap J$ and let $r \in R$. Then $ar \in I$ and $ar \in J$ so that $ar \in I \cap J$. $I \cup J$ is not an ideal.

9. Let $a \in I + J$ and $r \in R$. Then $a = i + j$ for some $i \in I, j \in J$. Then

$$ar = (i + j)r = ir + jr$$

with $ir \in I, jr \in J$.

11. Let I be an ideal in a ring R. Recall that $R/I = \{r + I | r \in R\}$ where addition is defined by

$$(r_1 + I) + (r_2 + I) = (r_1 + r_2) + I.$$

Similarly, multiplication in R/I is defined by

$$(r_1 + I)(r_2 + I) = r_1 r_2 + I.$$

See Exercise 3.2.10, which indicates why this multiplicative operation is well defined. One can similarly show that the additive operation is also well defined.

13. Clearly $I \subseteq R$. Let $r \in R$. Then $er = r \in I$ so $R \subseteq I$.

Exercises 3.3

1. Let $b, c \in f^{-1}(B_1)$ so that $f(b), f(c) \in B_1$. Since f is a ring homomorphism, we have

$$f(b + c) = f(b) + f(c) \in B_1,$$

so that $b + c \in f^{-1}(B_1)$. Similarly for multiplication, $f(bc) = f(b)f(c) \in B_1$.

To check the properties for $f^{-1}(B_1)$ to be an ideal, we need to do the same check for addition, but for multiplication, we need to allow $b \in f^{-1}(B_1)$ and $c \in A$. Then

$$f(bc) = f(b)f(c) \in B_1,$$

so $bc \in f^{-1}(B_1)$, since B_1 is now assumed to be an ideal.

3. Let $x \in (a + I)(b + I)$ so that

$$x = (a + i_1)(b + i_2) = ab + ai_2 + i_1 b + i_1 i_2$$

for some $i_1, i_2 \in I$. Since I is an ideal, the elements $ai_2, bi_1 = i_1 b, i_1 i_2 \in I$. Hence $x \in ab + I$.

5. Let $a \in K$ and let $r \in A$. Then

$$f(ar) = f(a)f(r) = 0f(r) = 0 \in K.$$

Exercises 3.4

1. The set $\{0, 3\}$ is an ideal in the ring N.

3. Let e be the unity element in a subring S of the integral domain R.

5. The main point to be checked so that the set S forms a subring of the ring $\mathbb{Z}_8$ of integers modulo 8 is that it is closed under both the operations of addition and multiplication modulo 8. To check that S is an ideal, notice that if $a \in S$ and $r \in R$, then $ar = 2kr$, for some integer k, is even modulo 8 so that S is an ideal.

7. The ring $\mathbb{Z}_n$ is an integral domain only if n is a prime, in which case the ring $\mathbb{Z}_n$ is a field. To see this, if $n = ab$ with $1 < a, b < n$, then in the ring $\mathbb{Z}_n$, $ab = 0$, so a and b are zero divisors, and so the ring $\mathbb{Z}_n$ is not an integral domain.

Chapter 4

Exercises 4.1

1. Let F be a field and let $a \in F, a \neq 0$. Assume there is an element $b \in F, b \neq 0$ so that $ab = 0$. Since every non-zero element in a field has a multiplicative inverse, we can multiply both sides by b^{-1} to obtain $abb^{-1} = 0$, so that $a = 0$, a contradiction.

3. Let $a(x) = a_n x^n + \cdots + a_i x^i$, where $a_i \neq 0$ and the coefficients of all lower powers of x are zero. Similarly, let $b(x) = b_m x^m + \cdots + b_j x^j$ where b_j is the smallest coefficient that is non-zero in the polynomial $b(x)$.

Assume that $a(x)b(x) = 0$. Consider the term $a_i b_j x^{i+j}$ in the product polynomial $a(x)b(x)$. Then we must have that $a_i b_j = 0$ with neither $a_i = 0$ nor $b_j = 0$. This is not possible, since from Exercise 4.1.1, a field cannot have any zero divisors.

5. If n is a prime p, the ring $\mathbb{Z}_p$ is a field since every non-zero element of $\mathbb{Z}_p$ has a multiplicative inverse.

If $\mathbb{Z}_n$ is a field, assume that $0 \equiv n \equiv ab \pmod{n}$ for positive integers a and b with $1 < a, b < n$. Then a has an inverse a^{-1} (since every non-zero $a \in \mathbb{Z}_n$ is relatively prime to n) so that $a^{-1}ab \equiv b \equiv 0 \pmod{n}$, a contradiction, so that n must be a prime.

7. Over the complex numbers

$$x^2 + 5 = (x + \sqrt{5}i)(x - \sqrt{5}i),$$

recalling that $i^2 = -1$.

9. $a \cdot 0 = a \cdot (0 + 0) = a \cdot 0 + a \cdot 0$. By adding the element $-a \cdot 0$ to both sides, we see that $0 = a \cdot 0$.

11. Over the field $\mathbb{Z}_2$, $x^4 + 1 = (x^2 + 1)^2 = (x + 1)^4$.

13. The elements $1, 2, 3, 4, 5, 6$ have multiplicative orders $1, 3, 6, 3, 6, 2$ in $\mathbb{Z}_7$.

15. A hint is given in the statement of the exercise.

17. Since the characteristic is 2, $2a = a + a = 0$ for all $a \in F$, and hence $a = -a$.

19. The ring $\mathbb{Z}$ of integers is an integral domain but not a field, since, for example, 2 does not have an inverse in $\mathbb{Z}$. Actually only 1 and -1 have multiplicative inverses (and each is its own inverse) in the $\mathbb{Z}$.

21. To show F is a field, it is easy to show that the set F is closed under both of the operations of addition and multiplication. For example,

$$\begin{pmatrix} 1 & 1 \\ 1 & 0 \end{pmatrix} \begin{pmatrix} 0 & 1 \\ 1 & 1 \end{pmatrix} = \begin{pmatrix} 1 & 0 \\ 0 & 1 \end{pmatrix}.$$

The characteristic is 2 since the sum of any of the matrices with itself gives the zero matrix.

23. We check that the relation R is symmetric, leaving the checks for reflexivity and transitivity to the reader. For symmetry, assume that $(r, s)R(t, u)$ so that $ru = st$. Then $ts = ur$ so that $(t, u)R(r, s)$.

25. Let F be a field and let D and E be subfields of F. Since D and E are themselves fields with the same operations as in the field F, it is easy to check that the intersection $D \cap E$ is also a field with the same operations as in the field F.

Yes, use induction on the number of subfields.

Chapter 5

Exercises 5.1

1. Yes, since $3125 = 5^5$ and the base 5 is a prime. The charactertistic is 5.

3. Yes, since $729 = 3^6$ and the base 3 is a prime.

Exercises 5.2

1. The addition and multiplication tables are

$+$	0	1	θ	$\theta + 1$
0	0	1	θ	$\theta + 1$
1	1	0	$\theta + 1$	θ
θ	θ	$\theta + 1$	0	1
$\theta + 1$	$\theta + 1$	θ	1	0

$\times$	0	1	θ	$\theta + 1$
0	0	0	0	0
1	0	1	θ	$\theta + 1$
θ	0	θ	θ	0
$\theta + 1$	0	$\theta + 1$	0	$\theta + 1$

This ring is not a field, because θ and $\theta + 1$ are zero divisors since their product is zero and neither element is the zero element.

The reader should compare these tables with those computed earlier in Chapter 4 for the field $\mathbb{F}_{2^2}$. The addition tables for this ring and for the field $\mathbb{F}_{2^2}$ are the same, but the multiplication tables are not the same.

3. Let α be a root of the primitive polynomial $x^4 + x + 1$ over the binary field. Since α is a primitive element, the non-zero elements of the field $\mathbb{F}_{2^4}$ can then be viewed as

$$1, \alpha, \alpha^2, \alpha^3, \alpha^4, \alpha^5, \alpha^6, \alpha^7, \alpha^8, \alpha^9, \alpha^{10}, \alpha^{11}, \alpha^{12}, \alpha^{13}, \alpha^{14}$$

whose multiplicative orders are

$$1, 15, 15, 5, 15, 3, 5, 15, 15, 5, 3, 15, 5, 15, 15.$$

5. Recall that the order of an element a in $\mathbb{Z}_p$ must divide $p - 1$. First check that 3 is a primitive element modulo 17 so it has order 16 modulo 17. If we list the non-zero elements modulo 17 as $1, 2, \ldots, 16$, their orders will turn out to be

$$1, 8, 16, 4, 16, 16, 16, 8, 8, 16, 16, 16, 4, 16, 8, 2.$$

Exercises 5.3

1. For any prime p and any positive integer n, the number of monic irreducible polynomials of degree n over the field $\mathbb{Z}_p$ is given by

$$N_p(n) = \frac{1}{n} \sum_{d \mid n} \mu(d) p^{n/d};$$

see Theorem 5.2. In our examples, for $p = 2$ these numbers turn out to be 1, 2, 3; for $p = 3$ they are 3, 8, 18; and for $p = 5$ they are 10, 40, 150.

3. The elements of $\mathbb{F}_{2^4}$ can be viewed as

$$0, 1, \alpha, \alpha + 1, \alpha^2, \alpha^2 + 1, \alpha^2 + \alpha, \alpha^2 + \alpha + 1$$

$$\alpha^3, \alpha^3 + 1, \alpha^3 + \alpha, \alpha^3 + \alpha + 1, \alpha^3 + \alpha^2, \alpha^3 + \alpha^2 + 1, \alpha^3 + \alpha^2 + \alpha, \alpha^3 + \alpha^2 + \alpha + 1.$$

As an illustration to construct the addition table, $(\alpha + 1) + (\alpha^3 + \alpha) = \alpha^3 + 1$ as a polynomial in α. For multiplication,

$$(\alpha + 1)(\alpha^3 + \alpha) = \alpha^4 + \alpha^2 + \alpha^3 + \alpha = \alpha^3 + \alpha^2 + 1$$

since $\alpha^4 = \alpha + 1$.

5. Assume that the field $\mathbb{F}_q$ has p^k elements where $k > 1$. Then k always has at least the two divisors 1 and k. Hence the field $\mathbb{F}_{p^k}$ has at least two subfields $\mathbb{Z}_p$ and $\mathbb{F}_{p^k}$.

7. The divisors of 10 are 1, 2, 5, and 10. Hence the subfields of the field $\mathbb{F}_{2^{10}}$ are the fields $\mathbb{Z}_2, \mathbb{F}_{2^2}, \mathbb{F}_{2^5}, \mathbb{F}_{2^{10}}$.

Exercises 5.4

1. Over the field $\mathbb{Z}_5$, $f(3) = 3^{214} + 3(3^{152}) + 2(3^{47}) + 2$. By Fermat's Theorem (Theorem 1.22), $3^4 \equiv 1 \pmod 5$ so that $3^{4k} \equiv 1 \pmod 5$ for any positive integer k. Thus in $\mathbb{Z}_5$

$$f(3) \equiv 3^2 + 3 + 2(3^3) + 2 \equiv 3 \pmod 5.$$

3. Let $f(x) = x^4 - 2x^2 - 3$. Then it is easy to see that $f(2) = f(3) = 0$ so that $x - 2$ and $x - 3$ are roots of the polynomial $f(x)$. Now use long division, with arithmetic being done modulo 5, to obtain

$$x^4 - 2x^2 - 3 = (x - 2)(x - 3)(x^2 + 2).$$

The reader should then check that the polynomial $x^2 + 2$ is irreducible over the field $\mathbb{Z}_5$ to complete the factorization.

5. If $f : \mathbb{F}_q \to \mathbb{F}_q$ then for each $a \in \mathbb{F}_q$, a can be mapped to any one of q elements in the field $\mathbb{F}_q$. Since there are q distinct elements in the finite field $\mathbb{F}_q$, the total number of functions $f : \mathbb{F}_q \to \mathbb{F}_q$ is q^q.

Actually these q^q functions can be represented by q^q polynomials with coefficients in $\mathbb{F}_q$. The polynomials may be given by the q^q polynomials of degree $< q$ with coefficients in $\mathbb{F}_q$.

7. Let $f(x) = x^2$ over the binary field $\mathbb{Z}_2$.

9. Assume that $f(x) = \sum_{i=0}^n a_i x^i$ where each $a_i \in \mathbb{F}_q$. Then

$$(f(x))^q = (\sum_{i=0}^n a_i x^i)^q = \sum_{i=0}^n a_i^q x^{iq} = \sum_{i=0}^n a_i (x^q)^i = f(x^q).$$

We note that $a_i^q = a_i$ since $a_i \in \mathbb{F}_q$.

Exercises 5.5

1. (i) $x, x + 1$

(ii) $ax + b, a \neq 0, b \in \mathbb{Z}_5$ which gives $4(5) = 20$ PPs

(iii) $a(x + b)^3 + c, a, \neq 0, b, c \in \mathbb{Z}_5$, which gives $4(5^2) = 100$ PPs for a total of $100 + 20 = 120 = 5!$ PPs over the field $\mathbb{Z}_5$

3. Let $p(x)$ be the given polynomial. Then we see that $p(0) = 1$ and $p(1) = p - 3 + 2 + 1 = p = 0$ in $\mathbb{Z}_5$. For $a \neq 0, 1$ we have

$$p(a) = a^{p-2} + \cdots + a + 1 + a = \frac{a^{p-1} - 1}{a - 1} + a = 0 + a = a.$$

Hence the polynomial $p(x)$ represents the transposition (01) since it maps 0 to 1, 1 to 0, and fixes all other elements.

Exercises 5.6

1. First construct the field $\mathbb{F}_{2^2}$ using θ as a root of the irreducible polynomial $x^2 + x + 1$ over the binary field $\mathbb{Z}_2$.

Use the polynomials $x + y, \theta x + y, (\theta + 1)x + y$ where θ is a root of the irreducible polynomial $x^2 + x + 1$ over the binary field. Label the rows (and columns) of a 4×4 square by the field elements $0, 1, \theta, \theta + 1$.

For a nonzero element a, place the field element $ax + y$ at the intersection of row x and column y of the a-th square. Check that these polynomials give three mutually orthogonal latin squares of order 4.

After some calculation, the squares

$$
\begin{array}{cccc}
0 & 1 & \theta & \theta + 1 \\
1 & 0 & \theta + 1 & \theta \\
\theta & \theta + 1 & 0 & 1 \\
\theta + 1 & \theta & 1 & 0
\end{array}
$$

$$
\begin{array}{cccc}
0 & 1 & \theta & \theta + 1 \\
\theta & \theta + 1 & 0 & 1 \\
\theta + 1 & \theta & 1 & 0 \\
1 & 0 & \theta + 1 & \theta
\end{array}
$$

$$
\begin{array}{cccc}
0 & 1 & \theta & \theta + 1 \\
\theta + 1 & \theta & 1 & 0 \\
1 & 0 & \theta + 1 & \theta \\
\theta & \theta + 1 & 0 & 1
\end{array}
$$

provide a complete set of three MOLS of order 4.

If the reader prefers to view this set of three MOLS of order 4 as consisting of the numbers $0, 1, 2, 3$, one could, in each of the three squares, replace $0, 1, \theta, \theta + 1$ by the numbers $0, 1, 2, 3$. Thus the values 0 and 1 would stay the same, but each θ would be replaced by a 2 and each $\theta + 1$ would be replaced by a 3. After making these substitutions, the resulting latin squares are

$$
\begin{array}{cccc}
0 & 1 & 2 & 3 \\
1 & 0 & 3 & 2 \\
2 & 3 & 0 & 1 \\
3 & 2 & 1 & 0
\end{array}
$$

$$
\begin{array}{cccc}
0 & 1 & 2 & 3 \\
2 & 3 & 0 & 1 \\
3 & 2 & 1 & 0 \\
1 & 0 & 3 & 2
\end{array}
$$

$$
\begin{array}{cccc}
0 & 1 & 2 & 3 \\
3 & 2 & 1 & 0 \\
1 & 0 & 3 & 2 \\
2 & 3 & 0 & 1
\end{array}
$$

These latin squares provide a complete set of three MOLS of order 4.

3. Consider the irreducible polynomial $p(x) = x^3 + x + 1$ over the field $\mathbb{Z}_2$ (check that this polynomial is indeed irreducible over $\mathbb{Z}_2$). Let θ be a root of $p(x)$. Label the rows (and columns) of an 8×8 square with the field elements of $\mathbb{F}_{2^3}$, which can be represented as

$$
0, 1, \theta, \theta + 1, \theta^2, \theta^2 + 1, \theta^2 + \theta, \theta^2 + \theta + 1.
$$

To construct 2 MOLS of order 8, use the polynomials $x + y$ and $\theta x + y$ to build the squares. We could actually construct 7 MOLS of order 8, by placing the field element $ax + y, a \in \mathbb{F}_8, a \neq 0$ at the intersection of row x and column y of the a-th square.

5. First construct 2 MOLS of order three by using the polynomials $x+y$ and $2x+y$ modulo 3. Then consider two of the three MOLS of order 4 constructed in Exercise 5.6.1. In the 2 MOLS of order 4, we have replaced the elements θ and $\theta + 1$ by the elements 2 and 3, so the 2 MOLS of order four are now each constructed, as above, with the elements 0, 1, 2, 3.

Now "glue together" the two MOLS of order 3 and two MOLS of order 4 to obtain latin squares of order 12. To form a latin square of order 12, use 12 ordered pairs (a, b) where $a = 0, 1, 2$ comes from a latin square of order 3 and $b = 0, 1, 2, 3$ comes from a latin square of order 4.

To illustrate this glueing procedure (which is mathematically known as the Kronecker product), we will consider a small example where we will start with a latin square of order two and a latin square of order 3, yielding a latin square of order 6.

Consider the latin square of order two given by

$$
\begin{array}{cc}
0 & 1 \\
1 & 0
\end{array}.
$$

Also consider the latin square of order 3 given by

$$
\begin{array}{ccc}
0 & 1 & 2 \\
1 & 2 & 0 \\
2 & 0 & 1
\end{array}.
$$

We now illustrate the glueing process that will yield a latin square of order six. We will construct all $2 \times 3 = 6$ ordered pairs (a, b) where $a = 0, 1$ and $b = 0, 1, 2$. To simplify our notation, we will denote the ordered pair (a, b) by simply writing ab.

$$\begin{array}{cccccc}
00 & 01 & 02 & 10 & 11 & 12 \\
01 & 02 & 00 & 11 & 12 & 10 \\
02 & 00 & 01 & 12 & 10 & 11 \\
10 & 11 & 12 & 00 & 01 & 02 \\
11 & 12 & 10 & 01 & 02 & 00 \\
12 & 10 & 11 & 02 & 00 & 01
\end{array}$$

We note that if we glue together a latin square of order m and a latin square of order n, we will obtain a latin square of order mn, as illustrated in the above case when $m = 2$ and $n = 3$. In addition, if the squares A_1, A_2 are MOLS of order m, and B_1, B_2 are MOLS of order n, then after the glueing process, the squares A_1 with B_1 and A_2 with B_2, we will have a pair of MOLS of order mn.

7. First check that $g = 3$ is a primitive element modulo 17. Since $a = 3$ and $b = 5$, user A calculates $3^3 \equiv 10 \pmod{17}$, which is sent to B. User B then calculates $10^5 \equiv 6 \pmod{17}$.

Similarly, B calculates $3^5 \equiv 5 \pmod{17}$, which is sent to user A who then calcualtes $5^3 \equiv 6 \pmod{17}$. Hence users A and B now share the common key 6.

9. The secret key can be any integer a with $2 \leq a \leq p - 2$ as indicated in the discussion of the Diffie/Hellman scheme. Hence there are $p - 3$ choices.

Chapter 6

Exercises 6.1

1. Let $\frac{a}{b}$ be a rational number and let r be a real number. Then $\frac{a}{b}r = \frac{ar}{b}$ is a real number, so yes, the real numbers form a vector space over the field of rational numbers.

3. No, since if r is a real number that is irrational (for example, like the real numbers $\sqrt{2}$ or e or π) then $r\frac{a}{b} = \frac{ra}{b}$ is not a rational number since the numerator is not an integer.

Similarly for the complex numbers,

$$(c + di)\frac{a}{b} = \frac{ca}{b} + \frac{da}{b}i$$

is not a rational number if $d \neq 0$.

5. Yes, since

$$\frac{a}{b}\left(\frac{c}{d}x^2 + \frac{e}{f}x + \frac{g}{h}\right) = \frac{ac}{bd}x^2 + \frac{ae}{bf}x + \frac{ag}{bh}$$

is a polynomial with rational coefficients.

7. Since $a \in F$ is non-zero, it has a multiplicative inverse $a^{-1} \in F$. Multiplying both sides of the vector equation $a\mathbf{x} = a\mathbf{y}$ by a^{-1} yields the result.

9. The reader should check that by defining

$$(f + g)(x) = f(x) + g(x)$$

and

$$(\lambda f)(x) = \lambda f(x),$$

the conditions for a vector space are satisfied.

11. We prove Parts (a) and (b), leaving Parts (c) and (d) for the reader.

(a) For Part (a), $0\mathbf{x} = (0+0)\mathbf{x} = 0\mathbf{x} + 0\mathbf{x}$ so that upon adding the element $-0\mathbf{x}$ to both sides, we see that $0 = 0\mathbf{x}$.

(b) $\lambda\mathbf{0} = \lambda(\mathbf{0} + \mathbf{0}) = \lambda\mathbf{0} + \lambda\mathbf{0}$ so that upon adding the element $-\lambda\mathbf{0}$ to both sides, we see that $\mathbf{0} = \lambda\mathbf{0}$.

Exercises 6.2

1. Let a and b be real numbers. Assume that

$$a(2,3) + b(4,9) = (0,0).$$

Hence we have the system of equations

$$2a + 4b = 0$$

$$3a + 9b = 0.$$

The reader should check that this system of simultaneous linear equations has a unique solution $a = b = 0$, so the two vectors are linearly independent over the field of real numbers.

3. Check that a basis of the vector space of complex numbers over the field of real numbers is given by the vectors $1 = 1 + 0i$ and $i = 0 + i$. These vectors are linearly independent (check!) and they also span since

$$a + bi = a(1 + 0i) + b(0 + i).$$

Hence the dimension of the vector space of complex numbers over the field of real numbers is two.

5. These vectors are linearly dependent over the real numbers, since

$$-(2,3,4) + 2(3,4,5) - (4,5,6) = (0,0,0).$$

7. Let $a, b, c \in F$ and assume that

$$a\mathbf{u} = \mathbf{0}$$

$$b\mathbf{u} + b\mathbf{v} = \mathbf{0}$$

$$c\mathbf{u} + c\mathbf{v} + c\mathbf{w} = \mathbf{0}.$$

The first equation implies that $a = 0$. From the second equation, if $b \neq 0$ then $\mathbf{u} = -\mathbf{v}$, which is a contradiction since $\mathbf{u}$ and $\mathbf{v}$ are linearly independent over the field F.

Similarly, the reader should check that the third equation then implies that $c = 0$, so the three vectors are indeed linearly independent over the field F.

9. If $\mathbf{u}$ and $\mathbf{v}$ are linearly dependent, then there are scalars a and b in the field F, not both zero, so that $a\mathbf{u} + b\mathbf{v} = 0$, and hence if $a \neq 0$, then $\mathbf{u} = \frac{-b}{a}\mathbf{v}$, so the vector $\mathbf{u}$ is a scalar multiple of the vector $\mathbf{v}$.

Similarly if $\mathbf{u} = a\mathbf{v}$ for some scalar a, then $\mathbf{u} - a\mathbf{v} = 0$ so that $\mathbf{u}$ and $\mathbf{v}$ are linearly dependent over the field F.

11. Let a and b be non-zero real numbers. Consider the vectors $a(1,0) = (a,0)$ and $b(0,1) = (0,b)$. The reader should check that these two vectors are linearly independent over the field of real numbers.

13. Assume that $\mathbf{x}, \mathbf{y}, \mathbf{z}$ are linearly dependent. Consider the vector equation

$$c(\mathbf{x} + a\mathbf{y} + b\mathbf{z}) + d\mathbf{y} + e\mathbf{z} = c\mathbf{x} + (ca + d)\mathbf{y} + (cb + e)\mathbf{z} = \mathbf{0}.$$

If the vectors $\mathbf{x} + a\mathbf{y} + b\mathbf{z}, \mathbf{y}, \mathbf{z}$ are linearly independent, the above coefficients must all be 0 so that $c = d = e = 0$, a contradiction of the fact that $\mathbf{x}, \mathbf{y}, \mathbf{z}$ are linearly dependent.

A similar argument shows that if $\mathbf{x}, \mathbf{y}, \mathbf{z}$ are linearly dependent, then the vectors $\mathbf{x} + a\mathbf{y} + b\mathbf{z}, \mathbf{y}, \mathbf{z}$ must be dependent.

15. Let $\mathbf{v}_1, \ldots, \mathbf{v}_{n+1}$ be $n + 1$ vectors in a vector space of dimension n. Let $\mathbf{u}_1, \ldots, \mathbf{u}_n$ be a basis of the vector space. Then each $\mathbf{v}_i$ can be written as a linear combination of the basis vectors. This gives an $(n + 1) \times n$ system of linear equations. Since the system has n columns, the rank cannot be more than n. Hence the vectors $\mathbf{v}_1, \ldots, \mathbf{v}_{n+1}$ are linearly dependent.

17. To show that V is a vector space, one just needs to verify that each of the conditions for a vector space holds. For a basis, consider the matrices

$$\begin{pmatrix} 1 & 0 \\ 0 & 0 \end{pmatrix}, \begin{pmatrix} 0 & 1 \\ 1 & 0 \end{pmatrix}, \begin{pmatrix} 0 & 0 \\ 0 & 1 \end{pmatrix}.$$

Show that these three vectors (matrices) are linearly independent, they span, and hence the dimension of the vector space will be 3.

19. Note that the matrix $B = \lambda A$ so B is a scalar multiple of the matrix A. Hence by Exercise 6.2.9 the two vectors are linearly dependent over the field F.

21. Consider the vector equation

$$a(1,2) + b(4,3) = (a + 4b, 2a + 3b) = (0,0).$$

This leads to the pair of simultaneous equations

$$a + 4b = 0$$

$$2a + 3b = 0,$$

which upon multiplying the first equation by 2, shows that $5b = 0$. Hence in any field of characteristic not 5, b must be 0.

Exercises 6.3

1. It is easy to check that

$$(2y + 3z, y, z) + (2w + 3v, w, v) = (2(y + w) + 3(z + v), y + w, z + v)$$

is also in the set S. Similarly

$$\lambda(2y + 3z, y, z) = (2\lambda y + 3\lambda z, \lambda y, \lambda z) \in S.$$

A basis for S is given by the vectors $(2, 1, 0)$ and $(3, 0, 1)$.

3. If $x_1^2 + y_1^2 = z_1^2$ and $x_2^2 + y_2^2 = z_2^2$, then

$$x_1^2 + x_2^2 + y_1^2 + y_2^2 = z_1^2 + z_2^2$$

and

$$\lambda x_1^2 + \lambda y_1^2 = \lambda(x_1^2 + y_1^2) = \lambda z_1^2,$$

so both the operations of vector addition and scalar multiplication are closed. Thus the dimension of the vector space is 2 since the vectors $(0, 1, 1)$ and $(1, 0, 1)$ form a basis.

5. Let $H = \{ax | a \in F\}$ for a fixed vector $\mathbf{x}$ in the vector space V. If $a\mathbf{x}, b\mathbf{x} \in H$ then

$$a\mathbf{x} + b\mathbf{x} = (a + b)\mathbf{x} \in H,$$

so H is closed under the operation addition of vectors. Similarly

$$c(a\mathbf{x}) = (ca)\mathbf{x} \in H,$$

so H is closed under scalar multiplication, and so H is a subspace of the vector space V.

7. The vector space V and the single vector $\{0\}$ are always subspaces of any vector space V over a field.

Chapter 7

Exercises 7.1

1. Note that modulo 5, $x^3 - 5x^2 + 2x - 4 = x^3 + 2x + 1$. Then check that $f(0) = 1, f(1) = f(3) = 4, f(2) = f(4) = 3$.

3. No, since the polynomial $x^2 + x + 1$ is irreducible over the field of two elements and this polynomial does not divide $x^4 + 1$ over $\mathbb{Z}_2$.

5. Consider the two polynomials $f(x) = x^2$ and $g(x) = x^4$.

Exercises 7.2

1. Over the binary field $\mathbb{Z}_2$, $x^3 - 1 = (x - 1)(x^2 + x + 1)$ since $x^2 + x + 1$ is irreducible over the field $\mathbb{Z}_2$.

Over the ternary field $\mathbb{Z}_3$, $x^3 - 1 = (x - 1)(x^2 + x + 1) = (x - 1)(x + 2)^2$.

3. Let $f(x) = x^3 - x^2 + x + 1$. Since $f(0) = f(2) = 1$ and $f(1) = 2$ modulo

3, there is no root in the field $\mathbb{Z}_3$, so the cubic polynomial is irreducible over the field $\mathbb{Z}_3$.

5. Over $\mathbb{Z}_2$, $x^4 + x^2 + 1 = (x^2 + x + 1)^2$ with $x^2 + x + 1$ being irreducible. Over $\mathbb{Z}_3$, 1 is a root of the polynomial, so $x - 1$ is a factor. Similarly 2 is a root, in fact 2 is a double root, so the polynomial factors as $(x - 1)^2(x - 2)^2$.

7. In a field of characteristic p a prime, expanding $(a + b)^p$ by the binomial theorem, we see that each of the coefficients except the first and last is divisible by p. Hence each of these coefficients is 0 modulo the prime p. Hence $(a + b)^p = a^p + b^p$ in the field.

Exercises 7.3

1. Over the complex numbers $x^2 + 5 = (x + \sqrt{5}i)(x - \sqrt{5}i)$ since $i^2 = -1$.

3. Let $f(x) = x^4 - 1$. Then $f(i) = i^4 - 1 = 1 - 1 = 0$ so $x - i$ is a factor of $f(x)$. Similarly $f(-i) = 0$ so $x + i$ is a factor. Hence by long division we have $x^4 - 1 = (x - i)(x + i)(x - 1)(x + 1)$ over the complex numbers.

Exercises 7.4

1. Let $f(x) = x^3 - 2x^2 - x + 2$. Then $f(0) = 2, f(1) = f(2) = 0, f(3) = 3, f(4) = 0$ so the only roots of $f(x)$ over the field $\mathbb{Z}_5$ are 1, 2, and 4.

3.
$$D_5(x, a) = x^5 - 5ax^3 + 5a^2x$$
$$D_6(x, a) = x^6 - 6ax^4 + 9a^2x^2 - 2a^3$$
$$D_7(x, a) = x^7 - 7ax^5 + 14a^2x^3 - 7a^3x$$
$$D_8(x, a) = x^8 - 8ax^6 + 20a^2x^4 - 16a^3x^2 + 2a^4$$

5. There are no real solutions to this equation since the discriminant $b^2 - 4ac = 1 - 4 = -3 < 0$.

7. Using the formulas from Section 7.4, one could check that $x_1 = 2, x_2 = 1, x_3 = -3$ are solutions.

As an alternative method, notice that these integer roots each divide the constant term of the polynomial, which is -6. Thus if we had known this fact (that any integer root must divide the constant term), we could have quickly tried the various divisors of 6 until we found a root, say a. Then $x - a$ would be a factor, so dividing the polynomial by $x - a$ via long division, we would be left with a quadratic polynomial, which we could quickly solve by using the quadratic formula to obtain the other two roots. The difficulty is of course that in the beginning, we do not know if this polynomial has any integer roots.

9. The reader should check using the formulas from the text that $x_1 = -3$ is a root so $x + 3$ is a factor. Then after using long division we will obtain a quadratic polynomial whose roots will turn out to be $x_2 = 3, x_3 = 4$. If we expected the roots to involve integers, then as above we could just check each of the divisors, say d, of 36, and use long division by the various factors $x - d$ until we locate a root. Then the remaining quadratic polynomial equation can be solved using the quadratic formula.

11. Over the binary field $\mathbb{Z}_2$ we have

$$x^3 - 6x^2 + 11x - 6 = x^3 + x = x(x^2 + 1) = x(x + 1)^2.$$

13. Over the field of complex numbers

$$x^4 - 1 = (x^2 - 1)(x^2 + 1) = (x - 1)(x + 1)(x - i)(x + i).$$

15. The only possible integer roots must be divisors of 2. We note that $\pm 1, \pm 2$ are not roots, so the polynomial $x^7 - 8x + 2$ does not have an integer root.

17. If $\frac{a}{b}$ is a rational root then a must divide 2 and b must divide 1. Hence the only possible rational roots are $\pm 1, \pm 2$. None of these satisfy the polynomial equation, so the polynomial does not have any rational root.

Chapter 8

Exercises 8.6

1. The length is 4, the dimension is 2, and the minimum distance is 2.

The code contains $2^2 = 4$ codewords, which are then written as four tuples: $0000, 1001, 0110, 1111$.

3. The coset leaders are as follows:

$$Coset Leaders$$
$$0000$$
$$0001$$
$$0010$$
$$0011$$

The standard array can be listed as follows:

$$Standard Array$$

0000	1001	0110	1111
0001	1000	0111	1110
0010	1011	0100	1101
0011	1010	0101	1100

The syndromes are of the form $\mathbf{c}H$ where the vector $\mathbf{c}$ is a coset leader and the matrix H is parity check matrix, which can be assumed to be

$$H = \begin{pmatrix} 10 \\ 01 \\ 01 \\ 10 \end{pmatrix}.$$

The resulting syndromes are thus $00, 10, 01, 11$.

5. The parity check matrix H for the code from Exercise 8.6.3 is $H = (-A^T : I_{n-k})$, where $G = (I_k : A)$. Hence

$$\begin{pmatrix} 1 & 1 & 0 & 1 & 0 & 0 & 0 \\ 1 & 0 & 1 & 0 & 1 & 0 & 0 \\ 0 & 1 & 1 & 0 & 0 & 1 & 0 \\ 1 & 1 & 1 & 0 & 0 & 0 & 1 \end{pmatrix}.$$

The codewords of the code C can be obtained as all binary linear combinations of the rows of G, so the number of codewords is $2^3 = 8$.

The codewords will be listed as the first row in the standard array for the code C. The coset leaders can be listed as the first vector in each row of the standard array. Since this code has minimum distance 3, it can correct only one error. Thus we will list only the coset leaders whose weights are at most one. Hence we will provide the top 8 rows of the standard array, with the remaining 8 rows arising from coset leaders of weights greater than 2.

Hence the standard array is

0000000	1001101	0101011	0010111	1100110	1011010	0111100	1110000
1000000	0001101	1101011	1010111	0100110	0011010	1111100	0110000
0100000	1101101	0001011	0110111	1000110	1111010	0011100	1010000
0010000	1011101	0111011	0000111	1110110	1001010	0101100	1100000
0001000	1000101	0100011	0011111	1101110	1010010	0110100	1111000
0000100	1001001	0101111	0010011	1100010	1011110	0111000	1110100
0000010	1001111	0101001	0010101	1100100	1011000	0111110	1110010
0000001	1001100	0101010	0010110	1100111	1011011	0111101	1110001

Finally, the syndromes are obtained as the vectors $\mathbf{y}H^T$ where $\mathbf{y}$ is a coset leader. Here again, since the code has minimum distance 3 and hence can only correct one error, we list only the syndromes arising from coset leaders of weight at most 1. These syndromes are the 1×4 binary vectors

$$0000$$
$$1101$$
$$1011$$
$$0111$$
$$1000$$
$$0100$$
$$0010$$
$$0001$$

7. We must list all binary vectors of length 5 that are at a distance at most 2 from the vector 11011. We see from Exercise 8.6.6 that there should be 16 vectors in the sphere of radius 2 about any vector.

The vector 11011 is at distance 0 from 11011.

The following five vectors are at distance 1 from 11011:

$$01011, 10011, 11111, 11001, 11010.$$

The following 10 vectors are at distance 2 from 11011:

$$00011, 01111, 01001, 01010, 10111, 10001, 10010, 11101, 11110, 11000$$

9. Let $G = \begin{pmatrix} 1 & 0 & 1 & 1 \\ 0 & 1 & 1 & 1 \end{pmatrix}$ so the minimum weight of either row is 3, but the codeword $1011 + 0111 = 1100$, which is the sum of the two rows of G, has weight 2.

11. The possible messages are all the 8 binary triples, i.e., $000, \ldots, 111$. We illustrate the two methods with the message 011 so that $(011)G = (0111100)$ using the matrix method.

Using the row space methods we note that the above codeword is obtained as $R_2 + R_3$, where R_2 and R_3 denote rows 2 and 3 of the generator matrix G.

13. Permuting the rows of the generator matrix G will result in the same row space and hence the same codewords, so the two codes are certainly equivalent. Similarly permuting the columns of the generator matrix simply

interchanges the coordinates of each of the codewords, so the resulting code is equivalent.

Determine the codewords of the codes C_1 and C_2 by, for example, determining the row spaces of the generator matrices G_1 and G_2. Then in each of the codewords of C_1, interchange the elements in the four coordinates using the permutation $(13)(24)$, i.e., interchange the elements in coordinates 1 and 3 and then 2 and 4. The resulting set of codewords will be those of the code C_2, so the two codes are equivalent.

Chapter 9

Exercises 9.1

1. Use induction on the positive integer n. Since $a_1 = 1$, $a_1 + 1 = 2$, which is a power of 2. Assume for $n = k$ that we have $a_k + 1 = 2^m$ for some positive integer m. Then

$$a_{k+1} + 1 = (2a_k + 1) + 1 = 2(a_k + 1) = 2(2^m) = 2^{m+1}$$

is a power of 2.

3. Clearly $3^{2(1)} - 1 = 8$. Assume for $n = k$ that $3^{2k} - 1 = 8m$ for some positive integer m. Then

$$3^{2(k+1)} - 1 = 3^{2k}3^2 - 1 = (3^{2k} - 1)9 + 8 = (8m)9 + 8 = 8(9m + 1)$$

and we are finished.

5. No, for example, $2^4 - 1 = 15$ is not a prime.

7. Use induction on the positive integer n. For $n = 1$ we have $1(2) = 1(2)(3)/3$. Assume now that for $n = k$ we have that

$$1(2) + 2(3) + \cdots + k(k + 1) = \frac{k(k + 1)(k + 2)}{3}.$$

Then for $n = k + 1$ we have

$$1(2) + 2(3) + \cdots + k(k+1) + (k+1)(k+2) = \frac{k(k + 1)(k + 2)}{3} + (k+1)(k+2),$$

which can be simplified to obtain $\frac{(k+1)(k+2)(k+3)}{3}$, which completes the proof.

Exercises 9.2

1. Clearly the integer ab is a common multiple of a and b. The $\gcd(a, b)$ is the largest common divisor of a and b, so $ab/\gcd(a, b)$ will be the least common multiple.

3. Assume that $(\frac{a}{b})^2 = k$ so that $a^2 = b^2k$. But then

$$\frac{a^2}{b^2}k = \left(\frac{a}{b}\right)^2 k,$$

a contradiction, since k would then be a square.

Exercises 9.3
    1. $A \cup B = \{a, b, 1, 2, c\}$; $A \cap B = \emptyset$
$A \times B = \{(a, 1), (a, 2), (a, c), (b, 1), (b, 2), (b, c)\}$
    3. The eight subsets are

$$\emptyset, \{1\}, \{3\}, \{5\}, \{1, 3\}, \{1, 5\}, \{3, 5\}, \{1, 3, 5\}$$

    There are 9 elements in the Cartesian product set. The elements of $A \times A$ are

$$\{(1, 1), (1, 3), (1, 5), (3, 1), (3, 3), (3, 5), (5, 1), (5, 3), (5, 5)\}.$$

    The number of elements in the power set $P(A \times A)$ is 2^9.
    5. Since for any integer a, $a - a = 0 = 2(0)$, the relation is reflexive. If $a - b = 2k$ for some integer k, then $b - a = 2(-k)$ is even, so the relation is symmetric.
    Finally if $a - b = 2k$ and $b - c = 2l$ for some integers k and l, then

$$a - c = a - b + b - c = 2k + 2l = 2(k + l)$$

is even, so the transitive property holds. Thus the relation is an equivalence relation.
    The equivalence class of an element a is the set $\{a + 2k \mid k$ an integer$\}$. Hence there are two equivalence classes; the even integers form one class and the odd integers form the other class.
    7. Given an element a, there may not be an element b so that aRb. Hence we cannot use the symmetric property.
    9. We note that $(A \cup B) \times C$ consists of the ordered pairs (a, c) where $a \in A$ or $a \in B$ and $c \in C$. Similarly $(A \times C) \cup (B \times C)$ consists of all of the pairs (x, y) with $x \in A$ and $y \in C$ or $x \in B$ and $y \in C$.

Exercises 9.4
    1. There are a total of 9 functions. One of them is the function defined by $f(a) = 2$ and $f(b) = e$. This example is 1-1. The total number of 1-1 functions $f : A \to B$ is $3(2) = 6$. The total number of functions is $3^2 = 9$. There are no onto functions since there are more elements in the set B than there are in the set A.
    3. See the solution to Exercise 9.5.1.
    5. See the solution to Exercise 9.4.1. In particular, the function f defined there is 1-1 but not onto.

Exercises 9.5

1. It is enough to show that the composition of two 1-1 and onto functions is again a 1-1 and onto function. Assume that

$$f(g(x_1)) = f(g(x_2)).$$

Then $g(x_1) = g(x_2)$ since f is 1-1, and thus $x_1 = x_2$ since g is 1-1. Therefore the composition $f(g(x))$ is also 1-1.

To show the composition is onto, let $z \in X$. Since f is onto, there is a $y \in X$ so that $f(y) = z$. Similarly since g is onto, there is an element $x \in X$ so that $g(x) = y$. Thus $f(g(x)) = f(y) = z$, so the composition is onto.

3. We denote the images of 3, 5, 7 for each of the $6 = 3!$ permutations by

$$
\begin{array}{ccc}
3 & 5 & 7 \\
3 & 7 & 5 \\
5 & 3 & 7 \\
5 & 7 & 3 \\
7 & 3 & 5 \\
7 & 5 & 3
\end{array}
$$

5. If $\sigma = (423)$, then

$$\sigma^4 = (\sigma^2)^2 = (243)^2 = (234) = \sigma.$$

7. As in Exercise 9.5.1, it is easy to show that the inverse of a permutation is both 1-1 and onto and hence is also a permutation.

Exercises 9.6

1.

$$A + B = \begin{pmatrix} -2 & 6 \\ 3 & 1 \end{pmatrix}, A - B = \begin{pmatrix} 4 & 2 \\ 1 & -7 \end{pmatrix}$$

$$AB = \begin{pmatrix} 1 & 18 \\ -9 & -8 \end{pmatrix}, BA = \begin{pmatrix} 1 & -20 \\ 9 & -8 \end{pmatrix}$$

3.

$$A^2 = \begin{pmatrix} -5 & -14 \\ 4 & -13 \end{pmatrix}, A^3 = A^2 A = \begin{pmatrix} -43 & 49 \\ -14 & -15 \end{pmatrix}$$

5. The matrix A is non-singular as is the matrix B.

7. The matrix A is non-singular so it has an inverse A^{-1}. To determine the matrix A^{-1} we may use the matrix equation

$$\begin{pmatrix} a + 3c & b + 3d \\ 2a + 4c & 2b + 4d \end{pmatrix} = \begin{pmatrix} 1 & 0 \\ 0 & 1 \end{pmatrix}.$$

Hence $b = -3d$. Similarly, $2a + 4c = 0$ so $a = -2c$, so that $2b + 4d = -2d = 1$. After some algebra we see that $a = -2, b = 3/2, c = 1, d = -1/2$. Thus

$$A^{-1} = \begin{pmatrix} -2 & 3/2 \\ 1 & -1/2 \end{pmatrix}.$$

Exercises 9.7

1. $(2 + 3i) + (6 - 2i) = (2 + 6) + (3 - 2)i = 8 + i$

3. The complex conjugate of $(2 - i)^3$ is

$$\overline{(2 - i)^3} = (2 + i)^3 = 2 + 11i$$

since $i^2 = -1$ and $i^3 = -i$. The reader should check that one could also first calculate $(2 - i)^3$ and then take the complex conjugate to obtain the same answer.

5. $(2 + 3i)(6 - 2i) = 12 + 14i - 6i^2 = 18 + 14i$

7. After a bit of algebra, $(2 - 4i)^3 = 188 + 48i$

9. We use the facts that $i^2 = -1$ and $i^3 = -i$ to obtain the sequence of values $i, -1, -i, 1, i, -1, -i, 1$.

Bibliography

[1] G.E. Andrews, *Number Theory*, Saunders, Philadelphia, 1971; reissued by Dover, 1995.

[2] H. Anton, *Elementary Linear Algebra*, Eighth Ed., Wiley, New York, 2000.

[3] R.C. Bose, On the application of the properties of Galois fields to the construction of hyper-Graeco-Latin squares, *Sankhyā* 3(1938), 323-338.

[4] L.N. Childs, *A Concrete Introduction to Higher Algebra*, Sec. Ed., Undergraduate Texts in Mathematics, Springer, New York, 1995; First softcover printing, 2000.

[5] A.A. Clarke, W.C. Waterhouse, J. Brinkhuis, and C. Greiter, *Disquisitiones Arithmeticae*, reprinted English version, Springer, NY, 1986.

[6] J. Dénes and A.D. Keedwell, *Latin Squares and Their Applications*, Academic Press, New York, 1974.

[7] L.E. Dickson, *History of the Theory of Numbers*, Vol. 1(1919), Vol. 2(1920), Vol. 3(1923), Carnegie Institute, Washington, D.C.

[8] W. Diffie and M.E. Hellman, New directions in cryptography, *IEEE Trans. Information Theory* IT-22(1976), no. 6, 644-654.

[9] J.B. Fraleigh, *A First Course in Abstract Algebra*, Addison-Wesley, Reading, MA, 1982.

[10] J.A. Gallian, *Contemporary Abstract Algebra*, D.C. Heath, Lexington, MA, 1994.

[11] E. Grosswald, *Topics from the Theory of Numbers*, Sec. Ed., Birkhäuser, Boston, 1984.

[12] R.W. Hamming, Error detecting and error correcting codes, *Bell Syst. Tech. J.* 29(1950), 147-160.

[13] T. Hansen and G.L. Mullen, Primitive polynomials over finite fields, *Math. Comp.* 59(1992), 639-643; Supplement S47-S50.

[14] R. Hill, *A First Course in Coding Theory*, Oxford Appl. Math. and Comp. Sci. Ser., Clarendon Press, Oxford, 1986; reprinted 2009.

[15] W.C. Huffman and V. Pless, *Fundamentals of Error-Correcting Codes*, Cambridge University Press, Cambridge, 2003.

[16] J.F. Humphreys and M.Y. Prest, *Numbers, Groups and Codes*, Sec. Ed., Cambridge University Press, Cambridge, 2004.

[17] T. Hungerford, *Abstract Algebra*, Saunders, Philadelphia, 1990.

[18] C.F. Laywine and G.L. Mullen, *Discrete Mathematics Using Latin Squares*, Wiley-Interscience Series in Discrete Mathematics and Optimization, John Wiley and Sons, Inc., New York, 1998.

[19] R. Lidl and G.L. Mullen, The world's most interesting class of integral polynomials, *J. Comb. Math. Comb. Comp.*, 37(2001), 87-100.

[20] R. Lidl, G.L. Mullen, and G. Turnwald, *Dickson Polynomials*, Longman Scientific and Technical, Essex, United Kingdom, 1993; now distributed by CRC Press, Boca Raton, FL.

[21] R. Lidl and H. Niederreiter, *Introduction to Finite Fields and Their Applications*, Rev. Ed., Cambridge University Press, Cambridge, 1994.

[22] R. Lidl and H. Niederreiter, *Finite Fields*, Rev. Ed., Cambridge University Press, Cambridge, 1997.

[23] F.J. MacWilliams and N.J.A. Sloane, *The Theory of Error-Correcting Codes*, Vol. 16, North Holland, Amsterdam, 1997, Eleventh impression, 1993.

[24] G.L. Mullen, A candidate for the "Next Fermat Problem," *Math. Intelligencer* 17(1995), 18-22.

[25] G.L. Mullen and C. Mummert, *Finite Fields and Applications*, Student Math. Library, Vol. 41, Amer. Math. Soc., Providence, RI, 2007.

[26] G.L. Mullen and D. Panario, *Handbook of Finite Fields*, CRC Press, Boca Raton, FL, 2013.

[27] G.L. Mullen and D. White, A polynomial representation for logarithms in $GF(q)$, *Acta Arithmetica* 47(1986), 255-261.

[28] I. Niven and H.S. Zuckerman, *An Introduction to the Theory of Numbers*, Fourth Ed., Wiley, New York, 1980.

[29] V. Pless, *Introduction to the Theory of Error-Correcting Codes*, Wiley-Interscience Series in Discrete Math., North Holland, New York, 1982.

[30] R.L. Rivest, A. Shamir, and L. Adelman, A method for obtaining digital signatures and public-key cryptosystems, *Comm. ACM* 21(1978), no. 2, 120-126.

[31] D.R. Stinson, A short proof of the nonexistence of a pair of orthogonal latin squares of order 6, *J. Combinatorial Theory, Ser.A*, 36(1984), 373-376.

[32] T. Sundstrom, *Mathematical Reasoning*, Prentice Hall (Pearson Education, Inc.), Upper Saddle River, NJ, 2003.

Index